T0321802

Gas Sweetening and Processing Field Manual

Gas Sweetening and Processing Field Manual

Maurice Stewart
Ken Arnold

AMSTERDAM • BOSTON • HEIDELBERG • LONDON • NEW YORK • OXFORD
PARIS • SAN DIEGO • SAN FRANCISCO • SINGAPORE • SYDNEY • TOKYO
Gulf Professional Publishing is an imprint of Elsevier

Gulf Professional Publishing is an imprint of Elsevier
225 Wyman Street, Waltham, MA 02451, USA
The Boulevard, Langford Lane, Kidlington, Oxford, OX5 1GB, UK

Notices
Knowledge and best practice in this field are constantly changing. As new
research and experience broaden our understanding, changes in research
methods, professional practices, or medical treatment may become necessary.
    Practitioners and researchers must always rely on their own experience
and knowledge in evaluating and using any information, methods,
compounds, or experiments described herein. In using such information or
methods they should be mindful of their own safety and the safety of others,
including parties for whom they have a professional responsibility.
    To the fullest extent of the law, neither the Publisher nor the authors,
contributors, or editors, assume any liability for any injury and/or damage
to persons or property as a matter of products liability, negligence or
otherwise, or from any use or operation of any methods, products,
instructions, or ideas contained in the material herein.

Library of Congress Cataloging-in-Publication Data
Application Submitted.

British Library Cataloguing-in-Publication Data
A catalogue record for this book is available from the British Library.

ISBN: 978-1-85617-982-9

For information on all Gulf Professional Publishing
publications visit our Web site at www.elsevierdirect.com

11   12   13   10 9 8 7 6 5 4 3 2 1

Printed and bound in the USA

Contents

Part 1
Gas Sweetening[1,2]

Contents

PROCESSING NATURAL GAS

Introduction

Natural gas used by consumers is significantly different from the natural gas that is brought up from the wellhead.

The processing of natural gas, in many respects, is less complicated than the processing of crude oil but is equally important before it is used by the end users.

Natural gas used by consumers is composed almost entirely of methane.

Natural gas found at the wellhead, although composed primarily of methane, contains a number of impurities that need to be removed.

Raw natural gas comes from three types of wells:

Oil wells

DOI: 10.1016/B978-1-85617-982-9.00002-8

Gas wells, and

Condensate wells.

Natural gas from oil wells

Termed "associated gas"

Gas can exist

Separate from oil in the formation and is termed "free gas" or

Dissolved in the crude oil and is termed "dissolved gas."

Natural gas from gas and condensate wells

Contain little or no crude oil

Termed "nonassociated gas"

Gas wells produce raw natural gas by itself

Condensate wells produce free natural gas along with a semiliquid hydrocarbon condensate

Whatever the source of raw natural gas

Exists in mixtures with

Other associated hydrocarbons such as

Ethane

Propane

Butane

i-Butane

Pentanes plus (natural gasoline)

Impurities

Water vapor

Hydrogen sulfide (H_2S)

Carbon dioxide (CO_2)

Helium

Nitrogen

Other compounds

Natural Gas Processing

Consists of separating all of the various hydrocarbons and impurities from the raw natural gas to produce what is termed "pipeline quality" dry natural gas.

Pipeline companies impose restrictions on the makeup of natural gas that is allowed into the pipeline.

Requires natural gas to be purified by removing

>Impurities termed "waste products" and

>Associated hydrocarbons such as

>>Ethane

>>Propane

>>Butane

>>*i*-Butane

>>Pentane plus (natural gasoline)

Associated hydrocarbons

>Termed "natural gas liquids" (NGLs)

>Can be very valuable by-products of natural gas processing

>Sold separately and have a variety of different uses such as

>>Enhancing oil recovery in oil wells

>>Providing raw materials for oil refineries and/or petrochemical plants, and

>>Sources of energy

Some of the needed processing can be accomplished at or near the wellhead (field processing).

Complete processing of natural gas takes place at a processing plant, usually located in a natural gas producing region.

Extracted natural gas is transported to processing plants through a network of gathering pipelines.

>Consists of small-diameter low pressure pipelines

>A complex gathering system can consist of thousands of miles of pipeline, interconnecting the processing plant to more than 100 wells.

In addition to processing done at the wellhead and centralized processing plants, some final processing is also done at "staddle extraction plants."

>Plants are located on major pipeline systems.

>Removes small quantities of NGLs that may still exist in pipeline quality gas

The processing of natural gas to meet pipeline quality gas involves the following four main processes to remove the various impurities

Oil and condensate removal

Water removal

Separation of NGLs

Sulfur and carbon dioxide removal

In addition the four processes above, heaters and scrubbers are installed, usually at or near the wellhead.

Scrubbers remove sand and other large-particle impurities.

Heaters ensure that the temperature of the gas does not drop below the hydrate formation temperature.

Hydrates

They have a tendency to form when the gas temperature drops.

Hydrates are solid or semisolid compounds, resembling ice-like crystals.

Should these hydrates accumulate, they can impede passage of natural gas through valves and gathering systems.

To reduce the occurrence of hydrates, the following equipment may be used:

Indirect fired heater

Hydrate inhibitors

Dehydration

Low temperature units

Oil and Condensate Removal

In order to process and transport associated dissolved natural gas

Gas must be separated from the oil in which it is dissolved

Often accomplished by using equipment installed at or near the wellhead.

Actual processes used to separate oil from natural gas and the equipment used can vary widely.

Dry pipeline quality natural gas is basically identical across different geographic regions.

Raw natural gas from different regions may have different compositions and separation requirements.

In some instances, natural gas is dissolved in oil underground primarily due to the pressure that the formation is under.

> When this natural gas and oil is produced, it is possible that it will separate on its own, simply due to decreased pressure (similar to opening a bottle of soda and allowing the release of dissolved carbon dioxide).

> Separation of oil and gas is relatively easy, and the two hydrocarbons are separated and exit the separator for further processing.

> Conventional separators are used which uses gravity separation to separate the heavy liquids (oil and water) and lighter fluid (natural gas).

In some instances, specialized process equipment, such as a low-temperature separator (LTX), is necessary to separate oil and natural gas.

> LTX is used for wells producing high pressure gas along with light crude oil or condensate.

> These separators use pressure differentials to cool the wet natural gas and separate the oil and condensate.

> Wet gas enters the separator, being cooled slightly by a heat exchanger.

> The gas then travels through a high pressure liquid "knockout" which serves to remove any liquids into a LTX.

> Gas then flows into this LTX through a choke mechanism, which expands the gas as it enters the separator.

> The rapid expansion of the gas allows for the lowering of the temperature in the separator.

> After liquid removal, the dry gas then travels back through the heat exchanger and is warmed by the incoming wet gas

> By varying the pressure of the gas in various sections of the separator, it is possible to vary

the temperature, which causes the oil and some water to be condensed out of the wet gas stream

This basic pressure–temperature relationship can work in reverse as well, to extract gas from a liquid oil stream.

Water Removal

In addition to separating oil and some condensate from the wet gas stream, it is necessary to remove most of the associated water.

Most of the liquid, free water associated with extracted natural gas, is removed by simple separation methods at or near the wellhead.

The removal of the water vapor that exists in solution in natural gas requires a more complex treatment.

This treatment consists of "dehydrating" the natural gas, which usually involves one of the following two processes:

Adsorption

Absorption

Adsorption

Occurs when the water vapor is taken out by a dehydrating agent.

Occurs when the water vapor is condensed and collected on the surface.

Glycol Dehydration

An example of absorption dehydration is glycol dehydration.

A liquid desiccant dehydrator serves to absorb water vapor from the gas stream.

The principle agent in this process is glycol which has a chemical affinity for water.

When glycol comes in contact with a stream of natural gas that contains water, the glycol absorbs the water vapor out of the stream.

Glycol dehydration involves using a glycol solution, usually either diethylene glycol (DEG) or triethylene glycol

(TEG), which is brought into contact with the wet gas stream in what is called a "contactor."

The glycol absorbs the water from the wet gas.

Once absorbed, the glycol particles become heavier and sink to the bottom of the contactor where they are removed.

The natural gas, having been stripped of most of its water content, is then transported out of the dehydrator.

The glycol solution, bearing all of the water stripped from the natural gas, is put through a specialized boiler designed to vaporize only the water out of the solution.

While water has a boiling point of 212 F (100 °C), glycol does not boil until 400 °F (204 C).

This boiling point differential makes it relatively easy to remove water from the glycol solution, allowing it to be reused in the dehydration process.

A new innovation in this process has been the addition of flash tank separator–condensers.

As well as absorbing water from the wet gas stream, the glycol solution occasionally carries with it small amounts of methane and other compounds found in the wet gas.

In the past, methane was simply vented out of the rebolier.

In addition to loosing a portion of the natural gas that was extracted, this venting contributes to air pollution and the greenhouse effect.

In order to decrease the amount of methane and other compounds that are lost, flash tank separator–condensers work to remove these compounds before the glycol solution reaches the boiler.

A flash tank separator consists of a device that reduces the pressure of the glycol solution stream, allowing the methane and other hydrocarbons to vaporize or "flash."

The glycol solution then travels to the reboiler, which may also be fitted with air or water cooled condensers, which serve to capture any remaining organic compounds that may remain in the glycol solution.

In practice, these systems recover 90–99% of methane that would otherwise be flared into the atmosphere.

Solid-Desiccant Dehydration

Solid-desiccant dehydration is the primary form of dehydrating natural gas using adsorption and usually consists of two or more adsorption towers, which are filled with a solid desiccant.

Typical desiccants include

Activated alumina

Silica gel

Molecular sieve

Wet natural gas is passed through these towers from top to bottom.

As the wet gas passes around the particles of desiccant material, water is retained on the surface of these desiccant particles.

Passing through the entire desiccant bed, almost all of the water is adsorbed onto the desiccant material, leaving the dry gas to exit the bottom of the tower.

Solid-desiccant dehydrators are typically more effective than glycol dehydrators and are usually installed.

Where very dry gas is required, such as upstream of a cryogenic expander, LPG, and LNG plants

As a type of straddle system along natural gas pipelines.

These types of dehydration systems are best suited for large volumes of gas under very high pressure and are thus usually located on a pipeline downstream of a compressor station.

Two or more towers are required due to the fact that after a certain period of use (typically 8 h), the desiccant in a particular tower becomes saturated with water.

To "regenerate" the desiccant, a high-temperature heater is used to heat gas to a very high temperature.

Passing this heated gas through a saturated desiccant bed vaporizes the water in the desiccant tower, leaving it dry and allowing for further natural gas dehydration.

Separation of NGLs

Natural gas coming directly from a well contains many NGLs that are commonly removed.

In most instances, NGLs have a higher value as separate products, and it is thus economical to remove them from the gas stream.

The removal of NGLs usually takes place in a relatively centralized processing plant and uses techniques similar to those used to dehydrate natural gas.

There are two basic steps to the treatment of NGLs in the natural gas stream:

> First, the liquids must be extracted from the natural gas.

> Second, these NGLs must be separated themselves, down to their base components.

NGL Extraction

The two principle techniques for removing NGLs from the natural gas stream are

> Absorption method and

> Cryogenic expander process.

These two processes account for around 90% of total NGLs production.

Absorption Method

The absorption method of NGL extraction is very similar to using absorption for dehydration.

The main difference is that, in NGL absorption, an absorbing oil is used as opposed to glycol.

> The absorbing oil has an "affinity" for NGLs in much the same manner as glycol has an affinity for water.

Before the oil has picked up any NGLs, it is brought into contact with the absorption oil.

As the natural gas is passed through an absorption tower, it is brought into contact with the absorption oil which soaks up a high proportion of the NGLs.

> The "rich" absorption oil, now containing NGLs, exists the absorption tower through the bottom.

> It is now a mixture of

>> Absorption oil

>> Propane

Butanes

Pentanes, and

Heavier hydrocarbons.

The rich oil is fed into lean oil stills, where the mixture is heated to a temperature above the boiling point of the NGLs, but below that of the oil.

This process allows recovery of around

75% of butanes

85–90% pentanes and heavier molecules from the natural gas stream.

This basic absorption process can be modified to improve its effectiveness, or to target the extraction of specific NGLs.

In the refrigerated oil absorption method

Lean oil is cooled through refrigeration

Propane recovery can be upward of 90%, and

40% of ethane can be extracted from the natural gas stream.

Extraction of the other heavier NGLs can be close to 100% using this process.

Cryogenic Expansion Process

Cryogenic processes are also used to extract NGLs from natural gas.

Absorption methods can extract almost all of the heavier NGLs.

Lighter hydrocarbons, such as ethane, are often more difficult to recover from natural gas stream.

In certain instances, it is economic to simply leave the lighter NGLs in the natural gas stream.

If it is economic to extract ethane and other lighter hydrocarbons, cryogenic processes are required for high recovery rates.

Cryogenic processes consist of dropping the temperature of the gas stream to around minus 120 °F.

There are a number of different ways of chilling the gas to these temperatures.

One of the most effective methods is the use of the turbo-expander process.

> Uses external refrigerants to cool the natural gas stream

> An expansion turbine is used to rapidly expand the chilled gases, which causes the temperature to drop significantly.

> The rapid drop in temperature

>> Condenses ethane and other hydrocarbons in the gas stream

>> Maintains methane in gaseous form.

Process

> Allows recovery of about 90–95% of the ethane originally

> Converts some of the energy released when the natural gas stream is expanded into recompressing the gaseous methane effluent, thus saving energy costs associated with extracting ethane.

The extraction of NGLs from the natural gas stream produces both

> Cleaner, purer natural gas, as well as the

> Valuable hydrocarbons that are the NGLs themselves.

NGL Fractionation

Once NGLs have been removed from the natural gas stream, they must be broken down into their base components to be useful.

> The mixed stream of different NGLs must be separated out.

> Process used to accomplish this task is called fractionation.

Fractionation

> Based on the different boiling points of the different hydrocarbons in the NGL stream

> Occurs in stages consisting of the boiling off of hydrocarbons one by one

> The name of a particular fractionator gives an idea as to its purpose, as it is conventionally named for the hydrocarbon that is boiled off.

The entire fractionation process is broken down into steps, starting with the removal of the lighter NGLs from the stream.

The particular fractionators are used in the following order:

Deethanizer

> This step separates the ethane from the NGL stream.

Depropanizer

> The next step separates the propane.

Debutanizer

> This step boils off the butanes, leaving the pentanes and heavier hydrocarbons in the NGLs.

Butane splitter or deisobutanizer

> This step separates the *i*- and *n*-butanes.

By proceeding from the lightest hydrocarbons to the heaviest, it is possible to separate the different NGLs reasonably easily.

Sulfur and Carbon Dioxide Removal

In addition to water, oil, and NGL removal, one of the most important parts of gas processing involves the removal of sulfur and carbon dioxide.

Natural gas from some wells may contain significant amounts of sulfur and carbon dioxide.

Natural gas with hydrogen sulfide (H_2S) and other sulfur products is called "sour gas"

Sour gas is undesirable because

> Sulfur compounds can be extremely harmful, even toxic, if one breathes it.

> Can also be extremely corrosive.

Sulfur that exists in the gas stream can be extracted and marketed on its own.

> 15% of USA's sulfur production is obtained from gas processing plants.

Gas Sweetening Plant

Sulfur exists in natural gas as hydrogen sulfide (H_2S), and the gas is usually considered sour if the hydrogen sulfide content exceeds 5.7 mg of H_2S per cubic meter of natural gas.

The process for removing hydrogen sulfide and carbon dioxide from a natural gas stream is referred to as "sweetening" the gas.

The primary process for sweetening natural gas is similar to processes of glycol dehydration and NGL absorption.

Amine solutions are used to remove the hydrogen sulfide and carbon dioxide.

Process is known simply as the "amine process" and is used in the majority of onshore gas sweetening operations.

Gas with hydrogen sulfide and/or carbon dioxide is run through a tower, which contains the amine solution.

This solution has an affinity for carbon dioxide and hydrogen sulfide, and absorbs these contaminants much like glycol absorbing water.

The two principle amine solutions used are

Monoethanolamine (MEA)

Diethanolamine (DEA)

Either of the above compounds, in liquid form, absorbs carbon dioxide and hydrogen sulfide from natural gas as it passes through.

The effluent gas is virtually free of carbon dioxide and hydrogen sulfide compounds.

Like the process for NGL extraction and glycol dehydration, the amine solution used can be regenerated (i.e., the absorbed sulfur is removed), allowing it to be reused to treat more gas.

Although many gas sweetening plants use the amine absorption process, it is also possible to use solid desiccants like iron sponges and gas permeation.

Extracted sulfur can be sold if reduced to its elemental form.

Elemental sulfur is a bright yellow powder like material and can be seen in large piles near the treatment plants.

In order to recover elemental sulfur from the gas processing plant, the sulfur containing discharge from a gas sweetening process must be further treated.

The process used to recover sulfur is known as the Claus process and involves using thermal and catalytic reactions to extract the chemical sulfur from the hydrogen sulfide solution.

The Claus process is usually able to recover 97% of the sulfur that has been removed from the natural gas stream.

Once the natural gas has been fully processed and is ready to be consumed, it must be transported from those areas that produce natural gas, to those areas that require it.

The remainder of this part discusses in detail the processes used to sweeten natural gas streams containing carbon dioxide and hydrogen sulfide.

ACID GAS CONSIDERATIONS

Carbon dioxide (CO_2), hydrogen sulfide (H_2S), and other sulfur compounds, such as mercaptans, are known as acid gases and may require complete or partial removal to meet contract specifications.

Acid Gases

H_2S combined with water forms sulfuric acid.

CO_2 combined with water forms carbonic acid.

Both are undesirable because they

Cause corrosion and

Reduce heating value and thus sales value.

H_2S is poisonous and may be lethal.

Table 1-1 shows physiological effects of H_2S in air.

Sour Gas

Defined as natural gas with H_2S and other sulfur compounds

Sweet Gas

Defined as natural gas without H_2S and other sulfur compounds

Table 1-1 Effects of H₂S concentrations in air

Percent by Volume	Parts per Million by Volume	Grains Per 100 Standard Cubic Feet[a]	Milligrams Per Cubic Meter[a]	Physiological Effects
			CONCENTRATIONS IN AIR	
0.00013	0.13	0.008	0.18	Obvious and unpleasant odor generally perceptible at 0.13 ppm and quite noticeable at 4.6 ppm. As the concentration increases, the sense of smell fatigues and the gas can no longer be detected by odor.
0.002	10	1.26	28.83	Acceptable ceiling concentration permitted by federal OSHA standards.
0.005	50	3.15	72.07	Acceptable maximum peak above the OSHA acceptable ceiling concentrations permitted once for 10 min per 8-h shift, if no other measurable exposure occurs.
0.01	100	6.30	144.14	Coughing, eye irritation, loss of sense of smell after 3–15 min. Altered respiration, pain in eyes, and drowsiness after 15–30 min, followed by throat irritation after 1 h. Prolonged exposure results in a gradual increase in the severity of these symptoms.
0.02	200	12.59	288.06	Kills sense of smell rapidly, burns eyes and throat.
0.05	500	31.49	720.49	Dizziness, loss of sense of reasoning and balance. Breathing problems in a few minutes. Victims need prompt artificial resuscitation.
0.07	700	44.08	1008.55	Unconscious quickly. Breathing will stop and deaths will result if not rescued promptly. Artificial resuscitation is needed.
0.10+	1000+	62.98	1440.98+	Unconsciousness at once. Permanent brain damage or death may result unless rescued promptly and given artificial resuscitation.

[a]Based on 1% hydrogen sulfide = 629.77 gr/100 SCF at 14.696 psia and 59 °F, or 101.325 kPa and 15 °C.

Gas Sales Contracts Limit Concentration of Acid Compounds

CO_2

2–4% for pipelines.

Lowers Btu content.

CO_2 is corrosive.

20 ppm for LNG plants.

H_2S

¼ grain sulfur per 100 scf (approximately 4 ppm).

0.0004% H_2S.

2 ppm for LNG plants.

H_2S is toxic.

H_2S is corrosive (refer to NACE MR-01-75).

Partial Pressure

Used as an indicator if treatment is required

Defined as

PP = (total pressure of system) (mol% of gas)

Where CO_2 is present with water, a partial pressure

Greater than 30 psia (207 kPa) would indicate CO_2 corrosion might be expected.

Below 15 psia (103 kPa) would indicate CO_2 corrosion would not normally be a problem, although inhibition may be required.

Factors that influence CO_2 corrosion are those directly related to solubility, that is temperature, pressure, and composition of the water.

Increased pressure increases solubility and

Increased temperature decreases solubility.

H_2S may cause sulfide stress cracking due to hydrogen embrittlement in certain metals.

H_2S partial pressure greater than 0.05 psia (0.34 kPa) necessitates treating.

NACE RP 0186

Recommends special metallurgy to guard against H_2S (Figure 1-1a and b)

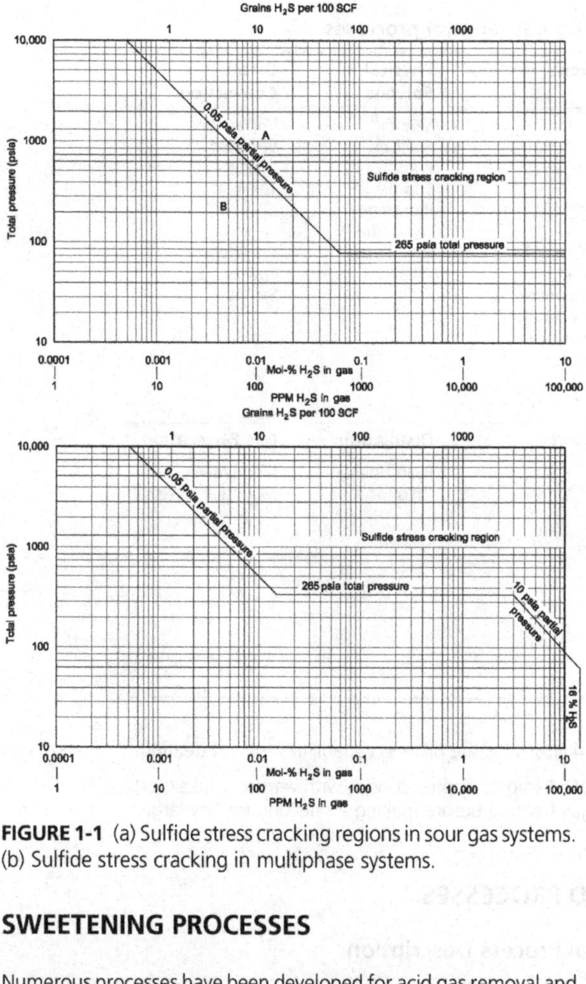

FIGURE 1-1 (a) Sulfide stress cracking regions in sour gas systems. (b) Sulfide stress cracking in multiphase systems.

SWEETENING PROCESSES

Numerous processes have been developed for acid gas removal and gas sweetening based on a variety of chemical and physical principles.

Table 1-2 lists the processes used to separate the acid gas from other natural gas components.

The list, although not complete, represents many of the common available commercial processes.

Table 1-3 shows the gases removed by various processes.

Table 1-2 Acid gas removal processes

Chemical Solvent	Physical Solvent	Direct Conversion
MEA	Selexol®	Iron sponge
DEA	Rectisol	Stretford
TEA	Purisol	Unisulf
MDEA	Spasolv	Takahax
DIPA/Shell ADIP®	Propylene carbonate	LO-CAT®
DGA/Fluor Econamine®	Estasolven	Lacy-Keller
Proprietary amine	Alkazid	Townsend
Benfield (hot carbonate)		Sulfint
Catacarb (hot carbonate)		
Giammarco-Vetrocoke (hot carbonate)		
Diamox		
Dravo/Still		

Specialty Solvent	Distillation	Gas Permeation
Sulfinol®	Ryan Holmes	Membrane
Amisol	Cryofrac	Molecular Sieve
Flexsorb PS		
Selefining		
Ucarsol LE 711		
Optisol		
Zinc oxide		
Sulfa-Check		
Slurrisweet		
Chemsweet		
Merox		

Table 1-4 illustrates the process capabilities for gas treating.

CAUTION: Designer should consult with vendors and experts in acid gas treating before making a selection for any large plant.

SOLID BED PROCESSES

General Process Description

Fixed bed of solid particles can be used to remove acid gases either through chemical reactions or through ionic bonding.

Gas stream flows through a fixed bed of solid particles which removes the acid gases and holds them in the bed.

When the bed is spent, the vessel must be removed from service and the bed regenerated or replaced.

Table 1-3 Gases removed by various processes

Process	GASES REMOVED				
	CO$_2$	H$_2$S	RHS	COS	CS$_2$
Solid Bed					
Iron sponge		X			
Sulfa-Treat		X			
Zinc Oxide		X			
Molecular Sieves	X	X	X	X	X
Chemical Solvents					
MEA—MonoEthanolAmine	X	X		X[a]	X
DEA—DiEthanolAmine	X	X		X	X
MDEA—MethylDiEthanolAmine		X			
DGA—DiGlycolAmine	X	X		X	X
DIPA—DiIsoPropanolAmine	X	X		X	
Hot potassium carbonate	X	X		X	X
Physical Solvents					
Fluor Solvent	X	X	X	X	X
Shell Sulfinol®	X	X	X	X	X
Selexol®	X	X	X	X	X
Rectisol		X			
Direct Conversion of H$_2$S to Sulfur					
Claus		X			
LO-CAT®		X			
SulFerox®		X			
Stretford		X			
Sulfa-Check		X			
Nash		X			
Gas Permeation	X	X			

[a]MEA reacts nonreversibly with COS (carbonyl sulfide), and, therefore, should not be used to treat gases with a large concentration of COS.

Since the bed must be removed from service to be regenerated, some spare capacity is normally provided.

Four commonly used processes under this category are

Iron sponge

Sulfa-Treat

Molecular sieve process

Zinc oxide process

Iron Sponge Process

Application

Economically applied to gases containing small amounts of H$_2$S (less than 300 ppm) operating at low to moderate pressures in the range of 50–500 psig (344.7 to 3447 kPa)

Table 1-4 Process capabilities for gas treating

	Normally Capable of Meeting 1/4 Grain[e] H_2S	Removes Mercaptans and COS Sulfur	Selective H_2S Removal	Solution Degraded (By)
Monoethanolamine	Yes	Partial	No	Yes (COS, CO_2, CS_2)
Diethanolamine	Yes	Partial	No	Some (COS, CO_2, CS_2)
Diglycolamine	Yes	Partial	No	Yes (COS, CO_2, CS_2)
Methyldiethanolamine	Yes	Slight	Yes[c]	No
Sulfinol®	Yes	Yes	Yes[c]	Some (CO_2, CS_2)
Selexol®	Yes	Slight	Yes[c]	No
Hot Pot-Benfield	Yes[a]	No[b]	No	No
Fluor Solvent	No[d]	No	No	–
Iron sponge	Yes	Partial	Yes	–
Mol Sieve	Yes	Yes	Yes[c]	CO_2 at high concentrations
Stretford	Yes	No	Yes	CO_2 at high concentrations
LO-CAT®	Yes	No	Yes	No
Chemsweet	Yes	Partial for COS	Yes	No

Source: GPSA Engineering Data Book, Tenth Edition, 1987.
[a]Hi-pure version.
[b]Hydrolizes COS only.
[c]Some selectivity is exhibited by these processes.
[d]Can be met with special design features.
[e]1/4 grain H_2S/100 scf ~ ppm H_2S.

Does not remove CO_2

Reaction of iron oxide and H_2S produces iron sulfide and water as follows:

$$Fe_2O_3 + 3H_2S \rightarrow Fe_2S_3 + 3H_2O$$
$$FeO + H_2S \rightarrow FeS + H_2O$$

Reaction requires the presence of slightly alkaline water (pH 8–10) and a temperature below 110 °F (47 °C).

> When temperatures exceed 110 °F (47 °C), careful control of pH must be maintained.
>
> If the gas does not contain sufficient water vapor, water may need to be injected into the inlet gas stream.
>
> The pH level can be maintained through the injection of caustic soda, soda ash, lime, or ammonia with the water.
>
> pH control should be avoided whenever possible.

Although the presence of free alkalines enhances H_2S removal, it

> Creates potential safety hazards
>
> Promotes formation of undesirable salts and
>
> Adds to capital costs.

Ferric oxide is impregnated on wood chips, which produce a solid bed with a large ferric oxide surface area.

> Several grades of treated wood chips are available, based on iron oxide content.
>
> Available in 6.5, 9.0, 15.0, and 20 lbs iron oxide/bushel

Chips are contained in a vessel, and sour gas flows downward through the bed and reacts with the ferric oxide.

Figure 1-2 shows a vertical vessel used in the iron sponge process.

Regeneration

Ferric sulfide can be oxidized with air to produce sulfur and regenerate the ferric oxide.

> Regeneration must be performed with care since the reaction with oxygen is exothermic (i.e., gives off heat).

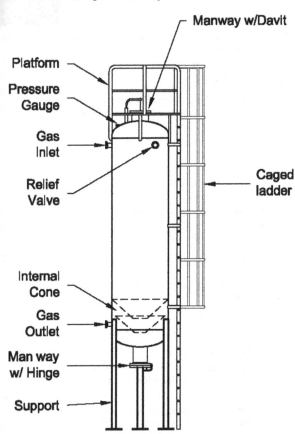

FIGURE 1-2 Iron oxide acid gas treating unit.

Air must be introduced slowly, so the heat of reaction can be dissipated.

If air is introduced quickly, the heat of reaction may ignite the bed.

For this reason, spent wood chips should be kept moist when removed from the vessel.

Otherwise, the reaction with oxygen in the air may ignite the chips and cause them to smolder.

The reactions for oxygen regeneration are as follows:

$$2Fe_2S_3 + 3O_2 + 2H_2O \rightarrow 2Fe_2O_3(H_2O) + 6S + Heat$$

$$4FeS + 3O_2 + 2xH_2O \rightarrow 2Fe_2O_3(H_2O)x + 4S + Heat$$

$$S_2 + 2O_2 \rightarrow 2SO_2$$

Some of the elemental sulfur produced in the regeneration step remains in the bed.

After several cycles, this sulfur will cake over the ferric oxide, decreasing the reactivity of the bed and causing excessive gas pressure drop.

Typically, after 10 cycles the bed must be removed from the vessel and replaced with a new bed.

Possible to operate an iron sponge with continuous regeneration by the introduction of small amounts of air in the sour gas feed.

The oxygen in the air regenerates the iron sulfide and produces elemental sulfur.

Although continuous regeneration decreases the amount of operating labor, it is not as effective as batch regeneration and it may create an explosive mixture of air and natural gas.

Due to the added costs associated with an air compressor, continuous regeneration generally does not prove to be the economic choice for the typically small quantities of gas involved.

Hydrate Considerations

Cooler operating temperatures of the natural gas, for example, during the winter, create the potential for hydrate formation in the iron sponge bed.

Hydrates can cause high-pressure drop, bed compaction, and flow channeling.

When the potential for hydrates exists, methanol can be injected to inhibit their formation.

If insufficient water is present to absorb the methanol, it may coat the bed, forming undesirable salts.

Hydrocarbon liquids in the gas tend to accumulate on the iron sponge media, thus inhibiting the reactions.

The use of a gas scrubber upstream of the iron sponge and a gas temperature slightly less than that of the sponge media may prevent significant quantities of liquids from condensing and fouling the bed.

There has been a recent revival in the use of iron sponges to sweeten light hydrocarbon liquids.

Sour liquids flow through the bed and are contacted with the iron sponge media and the reaction proceeds as above.

Sulfa-Treat Process

Process similar to iron sponge process utilizing the chemical reaction of ferric oxide with H_2S to sweeten gas streams.

Application

Economically applied to gases containing small amounts of H_2S

Carbon dioxide is not removed in the process.

Utilizes a proprietary iron oxide co-product mixed with inert powder to form a porous bed

Sour gas flows through the bed and forms a bed primarily of pyrite.

Powder has a bulk density of 70 lbs/ft^3 and ranges from 4–30 mesh.

Reaction works better with saturated gas and at elevated temperature up to 130 °F (54.4 °C).

No minimum moisture or pH level is required.

The amount of bed volume required increases as the velocity increases and as the bed height decreases.

Operation of the system below 40 °F (4.4 °C) is not recommended.

Beds are not regenerated and must be replaced when the bed is spent.

Molecular Sieve Process

Uses synthetically manufactured crystalline solids in a dry bed to remove gas impurities

> Crystalline structure of the solids provides a very porous material having uniform pore size.
>
> Within the pores the crystalline structure creates a large number of localized polar charges called active sites.
>
> Polar gas molecules such as H_2S and water vapor, which enter the pores, form weak ionic bonds at the active sites.
>
> Nonpolar molecules such as paraffin hydrocarbons will not bond to the active sites.
>
> Molecular sieves should be selected with a pore size that
>
> > Admits H_2S and water and
> >
> > Prevents heavy hydrocarbons and aromatic compounds from entering.
>
> Carbon dioxide molecules are about the same size as H_2S molecules, but are nonpolar.
>
> > CO_2 will enter the pores but will not bond to the active sites.
> >
> > Small quantities of CO_2 will be removed by becoming trapped in the pores by bonded H_2S or H_2O molecules blocking the pores.
> >
> > CO_2 will obstruct the access of H_2S and H_2O to the active sites, thus decreasing the overall effectiveness of the molecular sieve.
>
> Beds must be sized to remove all H_2O and provide for interference from other molecules in order to remove all H_2S.

Adsorption process usually occurs at moderate pressure.

Ionic bonds tend to achieve an optimum performance near 450 psig (3100 kPa), but can operate in a wide range of pressures.

Regeneration

> Regeneration is accomplished by flowing hot sweet stripping gas through the bed.
>
> Gas breaks the ionic bonds and removes the H_2S and H_2O.

Typical regeneration temperatures are in the range of 300–400 °F (150–200 °C).

Mechanical Degradation

Care should be taken to minimize mechanical damage to the solid crystals as this will decrease the bed's effectiveness.

Main cause is the sudden pressure and/or temperature changes that may occur when switching from adsorption to regeneration cycles.

Proper instrumentation can significantly extend bed life.

Application

Limited to small gas streams operating at moderate pressures

Generally used for polishing applications following one of the other processes

Zinc Oxide Process

Process

Equipment similar to the iron sponge process

Uses a solid bed of granular zinc oxide to react with the H_2S to form zinc sulfide and water as shown below

$$ZnO + H_2S \rightarrow ZnS + H_2O$$

Rate of reaction is controlled by the diffusion process, as the sulfide ion must first diffuse to the surface of the zinc oxide to react.

Temperatures above 250 °F (120 °C) increases the diffusion rate which promotes the reaction rate.

The strong dependence on diffusion means that other variables, such as pressure and gas velocity, have little effect on the reaction.

Bed Considerations

Contained in long thin beds to lessen the chances of channeling

Pressure drop through the beds is low.

Bed life is a function of gas H_2S content and can vary from 6 months to over 10 years.

Beds are often used in series to increase the level of saturation prior to change out of the catalyst.

Spent bed is discharged by gravity flow through the bottom of the vessel.

Application
Process seldom used due to disposal problems with the spent bed, which is classified as a heavy metal salt

CHEMICAL SOLVENT PROCESSES

General Process Description
Utilize an aqueous solution of a weak base to chemically react with and absorb the acid gases in the natural gas stream.

Absorption occurs as result of the partial pressure differential between the gas and the liquid phases.

Regeneration
Reactions are reversible by changing the system temperature or pressure, or both.

Aqueous base solution can be regenerated and circulated in a continuous cycle.

Most Common Chemical Solvents
Amines

Carbonates

Amine Processes

Amine Considerations
Several processes have been developed using the basic action of various amines.

Categorized by the number of organic groups bonded to the central nitrogen atom, as primary, secondary, or tertiary.

Primary amines form stronger bases than secondary amines, which form stronger bases than tertiary amines.

Amines with stronger base properties

Are more reactive toward CO_2 and H_2S gases and

Form stronger chemical bonds.

Implies that the acid gas vapor pressure will be lower for a given loading as amine reactivity increases and a higher equilibrium loading may be achieved

Process Description

A typical amine system is shown in Figure 1-3.

Sour gas enters the system through an inlet scrubber to remove any entrained water or hydrocarbon liquids.

Gas enters the bottom of the amine absorber and flows countercurrent to the amine solution.

Absorber tower consists of

Trays (diameters greater than 20 in. (500 mm))

Conventional packing (diameters less than 20 in. (500 mm))

Structured packing (diameters greater than 20 in. (500 mm))

Sweetened gas leaves the top of the tower.

Optional outlet scrubber may be included to recover entrained amine from the sweet gas.

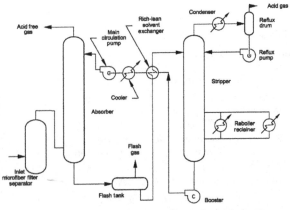

FIGURE 1-3 Gas sweetening process flow schematic (amine sweetening).

Since the natural gas leaving the top of the tower is saturated with water, the gas will require dehydration before entering a pipeline.

Rich amine, solution containing CO_2 and H_2S, leaves the bottom of the absorber and flows to the flash tank where

Most of the dissolved hydrocarbon gases or entrained hydrocarbon condensates are removed.

A small amount of the acid gases flashes to the vapor phase.

From the flash tank, the rich amine proceeds to the rich amine/lean amine heat exchanger.

Recovers some of the sensible heat from the lean amine stream which decreases the heat duty on the amine reboiler and the solvent cooler

Preheated rich amine then enters the amine stripping tower where heat from the reboiler breaks the bonds between the amine and acid gases.

Acid gases are removed overhead and lean amine is removed from the bottom of the stripper.

Hot lean amine flows to the rich amine/lean amine heat exchanger and then to additional coolers, typically aerial coolers, to lower its temperature to approximately 10 °F (5.5 °C) above the inlet gas temperature.

Reduces the amount of hydrocarbons condensed in the amine solution when the amine contacts the sour gas.

A side stream of amine, of approximately 3%, is taken off after the rich/lean amine heat exchanger and is flowed through a charcoal filter to clean the solution of contaminants.

Cooled lean amine is then pumped up to the absorber pressure and enters the top of the absorber.

Amine solution flows down the absorber where it absorbs the acid gases.

Rich amine is then removed at the bottom of the tower and the cycle is repeated.

Most common amine processes are

Monoethanolamine (MEA)

Diethanolamine (DEA)

Both processes remove CO_2 and H_2S to pipeline specifications.

Methyldiethanolamine (MDEA)

Newer process used for selective removal of H_2S in the presence of CO_2

Significantly reduces treating costs when CO_2 reduction is not necessary

Monoethanolamine (MEA) Systems

General Discussion

Primary amine, which has had widespread use as a gas sweetening agent

Process is well proven and can meet pipeline specifications.

MEA is a stable compound and in the absence of other chemicals suffers no degradation or decomposition at temperatures up to its normal boiling point.

Regeneration

Reactions are reversible by changing the system temperature.

Reactions with CO_2 and H_2S are reversed in the stripping column by heating the rich MEA to approximately 245 °F at 10 psig (118 °C at 69 kPa).

Acid gases evolve into the vapor and are removed from the still overhead.

Thus the MEA is regenerated.

Disadvantages

MEA reacts with carbonyl sulfide (COS) and carbon disulfide (CS_2) to form heat stable salts, which cannot be regenerated at normal stripping column temperatures.

At temperatures above 245 °F (118 °C) a side reaction with CO_2 exists which produces oxazolidone-2, a heat stable salt, which consumes MEA from the process.

Normal regeneration temperature in the still will not regenerate heat stable salts or oxazolidone-2.

Reclaimer is often included to remove these contaminants.

Reclaimer

A side stream of from 1% to 3% of the MEA circulation is drawn from the bottom of the stripping column.

Stream is then heated to boil the water and MEA overhead while the heat stable salts and oxazolidone-2 are retained in the reclaimer.

Reclaimer is periodically shut in and the collected contaminants are cleaned out.

When the contaminants are removed from the system, any MEA bonded to them is also lost.

Solution concentration and solution loading

Both are limited to avoid excessive corrosion.

MEA is usually circulated in a solution of 15–20% MEA by weight in water.

Operating experience indicates that the solution loading should not be greater than 0.3–0.4 mol of acid gas per mole of MEA.

Largely determined by the H_2S/CO_2 ratio.

The greater the ratio (i.e., the higher the concentration of H_2S relative to CO_2), the higher the allowable loading and amine concentration.

This is due to the reaction of H_2S and iron (Fe) to form iron sulfide (Fe_2S_3 and FeS), which forms a protective barrier on the steel surface.

This barrier can be stripped away by high fluid velocities and may lead to increased corrosion on the exposed steel.

Corrosion Considerations

Acid gases in the rich amine are corrosive, but the above concentration limits may hold corrosion to acceptable levels.

Corrosion commonly shows up:

Areas of carbon steel that have been stressed, such as heat affected zones near welds

Areas of high acid gas concentration, or

At a hot gas and liquid interface.

Thus, stress relieving all equipment after manufacturing is necessary to reduce corrosion, and special metallurgy is usually used in specific areas such as the still overhead or the reboiler tubes.

Foam Considerations

MEA systems foam rather easily resulting in excessive amine carryover from the absorber.

Foaming can be caused by a number of foreign materials such as

Condensed hydrocarbons

Degradation products

Solids such as carbon or iron sulfide

Excess corrosion inhibitor

Valve grease, etc.

Micro-fiber Filter Separator

Installed at the gas inlet to the MEA contactor

Effective method of foam control

Removes many of the contaminants before they enter the system

Hydrocarbon liquids are usually removed in the flash tank.

Degradation products are removed in a reclaimer as described above.

Blanket Gas System

Installed on MEA storage tanks and surge vessels

Prevents oxidation of MEA

Sweet natural gas or nitrogen are normally used.

MEA Losses

MEA has the lowest boiling point and the highest vapor pressure of the amines.

Results in MEA losses of 1–3 lbs/MMSCF ($16-48$ kg/MM m^3) of inlet gas

Summary

MEA systems can effectively treat sour gas to pipeline specifications.

Care in the design and material selection of MEA systems is required to minimize equipment corrosion.

Diethanolamine (DEA) Systems

General Discussion

Secondary amine is also used to treat natural gas to pipeline specifications.

As a secondary amine, DEA is less alkaline than MEA.

DEA systems do suffer the same corrosion problems, but not as severely as those using MEA.

Solution strengths are typically from 25% to 35% DEA by weight in water.

DEA has significant advantages over MEA when COS or CS_2 are present

DEA reacts with COS and CS_2 to form compounds which can be regenerated in the stripping column.

Thus, COS and CS_2 are removed without a loss of DEA.

Reclaimer

High CO_2 levels have been observed to cause DEA degradation to oxazolidones.

DEA systems usually include a carbon filter but do not typically include a reclaimer due to the small amount of degradation product.

Solution Concentration and Solution Loading

Stoichiometry reactions of DEA and MEA with CO and H_2S are the same.

Molecular weight of DEA is 105 compared to 61 for MEA.

Combination of molecular weights and reaction stoichiometry means that approximately 1.7 lbs (0.77 kg) of DEA must be circulated to react with the same amount of acid gas as 1.0 lbs (0.45 kg) of MEA

The solution strength of DEA ranges up to 35% by weight compared to 20% for MEA.

Loadings for DEA systems range from 0.35 to 0.65 mol of acid gas per mole of DEA without excessive corrosion.

The result of this is that the circulation rate of a DEA solution is slightly less than in a comparable MEA system.

Amine Losses

Vapor pressure of DEA is approximately 1/30 of the vapor pressure of MEA.

Thus, DEA amine losses are much lower than in an MEA system.

Diglycolamine® (DGA) Systems

General Discussion

DGA is a primary amine used in the Fluor Econamine Process to sweeten natural gas

Reactions of DGA with acid gases are the same as those for MEA

Unlike MEA, degradation products from reactions with COS and CS_2 can be regenerated.

Solution Concentration and Solution Loading

DGA systems typically circulate a solution of 50–70% DGA by weight in water.

At the above solution strengths and a loading of up to 0.3 mol of acid gas per mole of DGA, corrosion in DGA systems is slightly less than in MEA systems.

Advantages

Low vapor pressure decreases amine losses.

High solution strength permits lower circulation rates.

Diisopropanolamine (DIPA) Systems
General Discussion

DIPA is a secondary amine used in the Shell "ADIP®" process to sweeten natural gas.

Similar to DEA systems but offer the following advantages:

Carbonyl sulfide (COS) can be removed and the DIPA solution regenerated easily.

System is generally noncorrosive

Lower energy consumption

Advantages

At low pressures, DIPA will preferentially remove H_2S.

As pressure increases, the selectivity of the process decreases and DIPA removes increasing amounts of CO_2.

Thus, this system can be used either to selectively remove H_2S or to remove both CO_2 and H_2S.

Methyldiethanolamine (MDEA) Systems
General Discussion

Methyldiethanolamine (MDEA) is a tertiary amine, which like the other amines, is used to sweeten natural gas streams.

Major advantage over other amine processes:

MDEA selectivity for H_2S in the presence of CO_2

If the gas is contacted at pressures ranging from 800 to 1000 psig (5500–6900 kPa),

H_2S levels can be reduced to concentrations required by pipelines.

While at the same time, 40–60% of the CO_2 present flows through the contactor, untreated.

CO_2/H_2S Ratio

In cases where a high CO_2/H_2S ratio is present, MDEA can be used to improve the quality of the acid gas stream to a Claus recovery plant, but the higher CO_2 content of the treated residue gas must be tolerated.

Solution Concentration and Solution Loading

Solution strengths typically range from 40% to 50% MDEA by weight.

Acid gas loading varies from 0.2 to 0.4 or more moles of acid gas per mole of MDEA depending on supplier.

MDEA has a molecular weight of 119.

MDEA solution makeup is dependent upon the supplier. It can be adjusted to optimize treatment for a particular gas inlet composition.

Advantages

Higher allowable MDEA concentration and acid gas loading results in reduced circulation flow rates.

Significant capital savings are realized due to reduced pump and regeneration requirements.

MDEA has a lower heat requirement due to its low heat of regeneration.

In some applications, energy requirements for gas treating can be reduced as much as 75% by changing from DEA to MDEA.

Inhibited Amine Systems

General Discussion

Processes use standard amines that have been combined with special inhibiting agents which minimize corrosion.

Allows higher solution concentrations and higher acid gas loadings, thus reducing required circulation rates and energy requirements

Utilize hot potassium carbonate to remove CO_2 and H_2S

As a general rule, this process should be considered when the partial pressure of the acid gas is 20 psia (138 kPa) or greater.

Not recommended for low pressure absorption, or high pressure absorption of low concentration acid gas

Hot Potassium Carbonate Systems

General Discussion

Potassium bicarbonate ($KHCO_3$) solutions are not readily regenerable in the absence of CO_2.

Thus these processes are only employed for H_2S removal when quantities of CO_2 are present.

Potassium carbonate also reacts reversibly with COS and CS_2.

Process Description

Figure 1-4 shows a typical hot carbonate system for gas treating.

Gas to be treated enters the bottom of the absorber and flows countercurrent to the potassium carbonate.

Sweet gas exits the top of the absorber.

Absorber is typically operated at 230 °F (110 °C).

Gas/gas exchanger may be included to

    Cool the sweet gas

    Recover sensible heat, and

    Decrease the system's utility heat requirements.

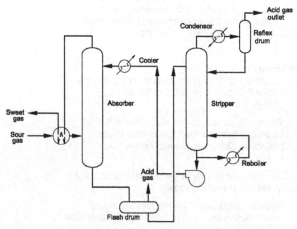

FIGURE 1-4 Gas sweetening flow schematic of a hot carbonate process.

Rich potassium carbonate solution from the bottom of the absorber flows to a flash drum where much of the acid gas is removed.

Solution then proceeds to the stripping column, which operates at approximately 245 °F (118 °C) and near atmospheric pressure.

Low pressure, combined with a small amount of heat input, strips the remaining acid gases.

Lean potassium carbonate from the stripper is pumped back to the absorber.

Lean solution may or may not be cooled slightly before entering the absorber.

Heat of reaction from the absorption of the acid gases causes a slight temperature rise in the absorber.

Solution concentration is limited by both

Solubility of potassium carbonate in the lean stream and

Solubility of the potassium bicarbonate ($KHCO_3$) in the rich stream.

Reaction with CO_2 produces two moles of $KHCO_3$ per mole of potassium carbonate reacted.

Thus, the $KHCO_3$ in the rich stream normally limits the lean solution potassium carbonate concentration to 20–35% by weight

Performance

Potassium carbonate works best on gas streams with a CO_2 partial pressure of 30–90 psi (207–620 kPa).

When CO_2 is not present, H_2S removal will be limited because the regeneration of the potassium carbonate requires an excess of $KHCO_3$.

The presence of CO_2 in the gas provides a surplus of $KHCO_3$ in the rich stream.

Pipeline quality gas often requires secondary treating using an amine or similar system to reduce H_2S level to 4 ppm.

Dead Spot Considerations

System is operated at high temperatures to increase the solubility of carbonates.

Thus, the designer must be careful to avoid dead spots in the system where the solution could cool and precipitate solids.

If solids do precipitate, the system may suffer from plugging, erosion, or foaming.

Corrosion Considerations

Hot potassium carbonate solutions are corrosive.

All carbon steel must be stress relieved to limit corrosion.

A variety of corrosion inhibitors, such as fatty amines or potassium dichromate, are available to decrease corrosion rates.

Proprietary Carbonate Systems

Several proprietary processes add a catalyst or activator.

Increase the performance of the hot potassium carbonate system

Increase the reaction rates both in the absorber and in the stripper

In general, these processes also decrease corrosion in the system.

Some of the proprietary processes for hot potassium carbonate include

Benfield:	Several activators
Girdler:	Alkanolamine activators
Catacarb:	Alkanolamine and/or borate activators
Giammarco-Vetrocoke:	Arsenic and other activator

Specialty Batch Chemical Solvents

General Discussion

Several batch chemical processes have been developed and have specific areas of application

Processes include

Zinc oxide slurry

Caustic wash

Sulfa-Check

Slurrisweet

Chemsweet

Process Description

Gas is flowed into a vessel and contacted with the solvent.

Acid components are converted to soluble salts, which are nonregenerable, limiting the life of the solution.

Once saturation levels are reached, the solution must be replaced.

For some of these processes, the spent solutions are not hazardous, but for others, the spent solutions have been labeled hazardous and, if used, must be disposed of as Class IV materials.

Performance

Units have a wide operating range, with acid gas concentrations ranging from as low as 10 ppm to as high as 20%.

Operating pressures range from near atmospheric to greater than 1000 psig (7000 kPa).

Units have been designed to handle from several thousand cubic feet per day to more than 15 MMSCFD (several hundred cubic meters per day to more than 420,000 m^3 per day).

Sulfa-Check

Sulfa-Check is a single step process that converts H_2S to sulfur in a bubble tower filled with a proprietary solution of oxidizing and buffering agents.

Oxidizing agent is a proprietary formulation of chelated nitrite ions.

Concentration Considerations

Reaction rate is independent of the concentration of the oxidizing agent.

There is no limit to the concentration of H_2S treated.

Process is most economical for acid gas streams containing from 1 ppm to 1% H_2S.

pH must be held above 7.5 to control selectivity and optimize H_2S removal.

One gallon (four liters) of oxidizing solution can remove up to 2 lbs (1 kg) of H_2S when the system is operated at ambient temperatures less than 100 °F (less than 38 °C).

If gas temperatures exceed 100 °F (38 °C), the solubility of sulfur in the oxidizing agent decreases.

Bubble Flow

Operating pressures of at least 20 psig (138 kPa) is required for proper unit operation to maintain bubble flow through the column.

Bubble flow is necessary to produce intimate mixing of the gas and liquid.

Disposal of Oxidizing Solution

Oxidizing solution will eventually become saturated and require replacement.

Disposal of this slurry poses no environmental problem, as the reaction produces an aqueous slurry of sulfur and sodium salt.

PHYSICAL SOLVENT PROCESSES

General Process Description

Similar to chemical solvent systems but are based on the gas solubility within a solvent instead of a chemical reaction.

Acid gas solubility depends on

Acid gas partial pressure and

System temperature.

Higher acid gas partial pressures increase the acid gas solubility.

Low temperatures increase acid gas solubility, but, in general, temperature is not as critical as pressure.

Various organic solvents are used to absorb the acid gases based on partial pressures.

Regeneration of the solvent is accomplished by

Flashing to lower pressures and/or

Stripping with solvent vapor or inert gas.

Some solvents can be regenerated by flashing only and require no heat.

Other solvents require stripping and some heat, but typically the heat requirements are small compared to chemical solvents.

Physical solvent processes have a high affinity for heavy hydrocarbons.

If the natural gas stream is rich in C_{3+} hydrocarbons, then the use of a physical solvent process may result in a significant loss of the heavier mole weight hydrocarbons.

These hydrocarbons are lost because they are released from the solvent with the acid gases and cannot be economically recovered.

Physical solvent processes should be considered for gas sweetening under the following circumstances:

Partial pressure of the acid gases in the feed is 50 psi (345 kPa) or higher

Concentration of heavy hydrocarbons in the feed is low

Only bulk removal of acid gases is required

Selective H_2S removal is required.

Figure 1-5 shows a physical solvent process.

Review of flow in Figure 1-5.

Sour gas contacts the solvent using countercurrent flow in the absorber.

Rich solvent from the absorber bottom is flashed in stages to near atmospheric pressure.

This causes the acid gas partial pressures to decrease, and the acid gases evolve to the vapor phase and are removed.

Regenerated solvent is then pumped back to the absorber.

Figure 1-5 is an example where flashing is sufficient to regenerate the solvent.

Some solvents require a stripping column just prior to the circulation pump.

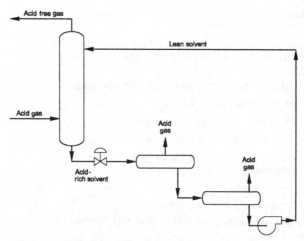

FIGURE 1-5 Typical flow schematic for a physical solvent process.

Some systems require temperatures below ambient, thus refrigeration using power turbines replaces the pressure reducing valves.

These turbines recover some of the power from the high pressure rich solvent and thus decrease the utility power requirements for refrigeration and circulation.

The majority of the physical solvent processes are proprietary and are licensed by the company that developed the process.

Four typical processes are discussed below.

Fluor Solvent Process

Uses propylene carbonate as a physical solvent to remove CO_2 and H_2S

Propylene carbonate also removes C_{3+} hydrocarbons, COS, SO_2, CS_2, and H_2O from the natural gas stream.

Thus, in one step the natural gas can be sweetened and dehydrated to pipeline quality.

Process is used for bulk removal of CO_2 and is not used to treat to less than 3% CO_2.

System requires special design features such as larger absorbers and higher circulation rates to obtain pipeline quality and usually is not economically applicable for these outlet requirements.

Propylene carbonate has the following characteristics, which make it suitable as a solvent for acid gas treating:

High degree of solubility for CO_2 and other gases

Low heat of solution for CO_2

Low vapor pressure at operating temperature

Low solubility for light hydrocarbons (C_1, C_2)

Chemically nonreactive toward all natural gas components

Low viscosity

Noncorrosive toward common metals

The above characteristics combine to yield a system that

Has low heat and pumping requirements.

Is relatively noncorrosive and

Suffers only minimal solvent losses, (less than 1 lbs/MMSCF)

Solvent temperatures below ambient are usually used to increase solvent gas capacity, and, therefore, decrease circulation rates.

Expansion of the rich solvent and flash gases through power turbines can provide the required refrigeration.

Alternately, auxiliary refrigeration may be included to further decrease circulation rates.

Sulfinol® Process

Developed and licensed by Shell

Employs both a chemical and a physical solvent for the removal of H_2S, CO_2, mercaptans, and COS

Sulfinol® solution is a mixture of

Tetrahydrothiophene dioxide (Sulfolane®) a physical solvent.

Diisopropanolamine (DIPA) a secondary amine, and Water.

DIPA, previously discussed, is the chemical solvent.

Solution concentrations range

> 25–40% Sulfolane®
>
> 40–55% DIPA, and
>
> 20–30% water
>
> Depending on the conditions and composition of the gas being treated.

Acid Gas Loadings

The presence of the physical solvent, Sulfolane®, allows higher acid gas loadings compared to systems based on amine only.

Typical loadings are 1.5 mol of acid gas per mole of Sulfinol® solution.

Higher acid gas loadings, together with a lower energy of regeneration, can result in lower capital and energy costs per unit of acid gas removed as compared to the ethanolamine processes.

Features of the Sulfinol® Process Include

Essentially complete removal of mercaptans

High removal rate of COS

Lower foaming tendency

Lower corrosion rates

Ability to slip up to 50% CO_2

Design Considerations

Design is similar to that of the ethanolamines.

Degradation of DIPA to oxazolidones (DIPA-OX) usually necessitates the installation of a reclaimer for their removal.

Foam Considerations

As with the ethanolamine processes, aromatics and heavy hydrocarbons in the feed gas should be removed prior to contact with the Sulfinol® solution to minimize foaming.

Factors to Consider Before Selecting a Treating Process

Merits of the Sulfinol® process as compared to the ethanolamine processes are many, but there are other factors which must be considered before selecting the appropriate gas treating process.

Licensing fees, while not necessary for the ethanolamine processes, is required for the Sulfinol® process.

Solvent costs are generally higher for Sulfinol® than they are for DEA.

Operators are more familiar with DEA and the typical problems associated with this process.

In cases of low acid gas partial pressure, the advantage of a lower circulation rate for the Sulfinol® process diminishes compared to DEA.

Selexol® Process

Uses the dimethylether of polyethylene glycol as a solvent.

Licensed by UOP.

Process is selective toward removing sulfur compounds.

Levels of CO_2 can be reduced by approximately 85%.

Process is economical when

High acid gas partial pressures exist and

An absence of heavy ends in the gas.

Process will not normally remove enough CO_2 to meet pipeline gas requirements.

DIPA can be added to the solution to remove CO_2 down to pipeline specifications.

Process also removes water to less than 7 lbs/MMSCF (0.11 g/stdm^3).

System then functions much like the Sulfinol® process discussed earlier.

The addition of DIPA increases the relatively low stripper heat duty.

Rectisol Process

Developed by the German Lurgi Company and Linde A. G.

Use methanol to sweeten natural gas.

Due to the high vapor pressure of methanol this process is usually operated at temperatures of 30 to −100 °F (−34 to −74 °C).

It has been applied for the purification of gas for LNG plants and in coal gasification plants, but is not commonly used to treat natural gas streams.

DIRECT CONVERSION PROCESSES

General Process Description

Chemical and physical solvent processes remove acid gas from the natural gas stream but release H_2S and CO_2 when the solvent is regenerated.

Release of H_2S to the atmosphere is limited by environmental regulations.

Acid gases could be routed to an incinerator/flare, which would convert the H_2S to SO_2.

Environmental regulations restrict the amount of SO_2 vented or flared.

Direct conversion processes use chemical reactions to oxidize H_2S and produce elemental sulfur.

Processes are generally based either on the reaction of H_2S and O_2 or H_2S and SO_2.

Both reactions yield water and elemental sulfur. These processes are licensed and involve specialized catalysts and/or solvents.

Stretford Process

General Discussion

Process uses O_2 to oxidize H_2S.

Originally licensed by the British Gas Corporation.

No longer in use.

Process Description

Figure 1-6 shows a simplified diagram of the process.

Gas stream is washed with an aqueous solution of sodium carbonate, sodium vanadate and anthraquinone disulfonic.

Oxidized solution is delivered from the pumping tank to the top of the absorber tower where it contacts the gas stream in a countercurrent flow.

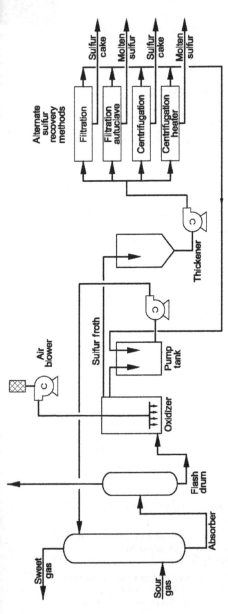

FIGURE 1-6 Simplified flow schematic of the stretford process.

The bottom of the absorber tower consists of a reaction tank from which the reduced solution passes to the solution flash drum, which is situated above the oxidizer. The reduced solution passes from here into the base of the oxidizer vessel.

Hydrocarbon gases, which have been dissolved in the solution at the plant pressure, are released from the top of the flash drum.

Air is blown into the oxidizer and the main body of the solution, now re-oxidized, passes into the pumping tank. The sulfur is carried to the top of the oxidizer by froth created by the aeration of the solution and passes into the thickener.

The function of the thickener is to increase the weight percent of sulfur which is pumped to one of the alternate sulfur recovery methods of filtration, filtration and autoclaves, centrifugation or centrifugation with heating.

Chemical reactions involved are

$$H_2S + Na_2CO_3 \rightarrow NaHS + NaHCO_3$$

Sodium carbonate provides the alkaline solution for initial adsorption of H_2S and the formation of hydrosulfide (HS).

The hydrosulfide is reduced in a reaction with sodium meta vandate to precipitate sulfur

$$HS + V^{+5} \rightarrow S + V^{(+4)}$$

Anthraquinone disulfonic acid (ADA) reacts with 4-valent vanadium and converts it back to 5-valent

$$V^{+4} + ADA \rightarrow V^{+5} + ADA(reduced)$$

Oxygen from the air converts the reduced ADA back to the oxidized state

$$Reduced\ ADA + O_2 \rightarrow ADA + H_2O$$

The overall reaction is

$$2H_2S + O_2 \rightarrow 2H_2O + 2S$$

IFP Process

General Discussion

Developed by the Institute Francais du Petrole

Process for reacting H_2S with SO_2 to produce water and sulfur

Overall reaction is

$$H_2S + SO_2 \rightarrow H_2O + 2S$$

Process Description

Figure 1-7 shows a simplified diagram of the IFP process.

Process involves mixing the H_2S and SO_2 gases and then contacting them with a liquid catalyst in a packed tower.

Elemental sulfur is recovered in the bottom of the tower.

A portion of this must be burned to produce the SO_2 required to remove the H_2S.

Ratio of H_2S to SO_2

The most important variable is the ratio of H_2S to SO_2 in the feed.

Controlled by analyzer equipment to maintain the system performance

LO-CAT®

General Discussion

Developed by ARI Technologies (LO-CAT) and Shell Development (Sulferox) combined development now under Merichem Company and marketed as LO-CAT

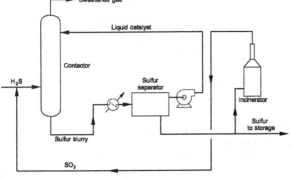

FIGURE 1-7 Simplified flow schematic of the IFP process.

Processes employ high iron concentration reduction-oxidation technology for the selective removal of H_2S (not reactive to CO_2) to less than 4 ppm in both low and high pressure gas streams.

Process Description

Acid gas stream is contacted with the solution where H_2S reacts with and reduces the chelated-iron and produces elemental sulfur.

The iron is then regenerated by bubbling air through the solution.

Heat is not required for regeneration.

The reactions involved are exothermic (gives off heat):

Absorption/reduction : $2Fe^{3+} + H_2S \rightarrow 2Fe^{2+} + S + 2H$

Regeneration/oxidation : $2Fe^{2+} + \frac{1}{2}O_2 + 2H \rightarrow 2Fe^{3+} + H_2O$

Overal chemistry : $H_2S + \frac{1}{2}O_2 \rightarrow S + H_2O$

Figures 1-8 and 1-9 shows a flow schematic for the LO-CAT® process

Operating Considerations

Turndown is 100%.

Solution is nontoxic; thus no special disposal problems.

No sulfur products dispersed to the atmosphere

Requires centrifuge and slurry handling

Sulferox®

Uses a patented pipeline contactor with co-current flow to minimize sulfur plugging

Claus

General Discussion

Used to treat gas streams containing high (above 50%) concentrations of H_2S

Chemistry of the units involves partial oxidation of hydrogen sulfide to sulfur dioxide, and the catalytically promoted reaction of H_2S and SO_2 to produce elemental sulfur.

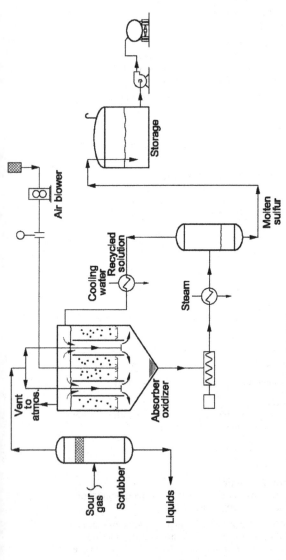

FIGURE 1-8 Simplified flow schematic of the LO-CAT process.

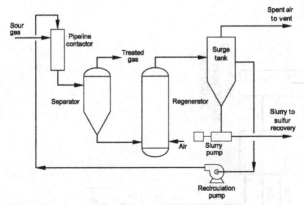

FIGURE 1-9 Simplified flow schematic of the sulferox process.

Reactions are staged and are as follows:

$H_2S + 3/2O_2 \rightarrow SO_2 + H_2O$ thermal state
$SO_2 + 2H_2S \rightarrow 3S + 2H_2O$ thermal and catalytic
    stage

Process Description

Figure 1-10 shows a simplified flow diagram of a two-stage Claus process plant.

First stage of the process converts H_2S to sulfur dioxide and to sulfur by burning the acid gas stream with air in the reaction furnace.

    This provides SO_2 for the next phase of the reaction.

Gases leaving the furnace are cooled to separate out elemental sulfur formed in the thermal stage.

Reheating, catalytically reacting and sulfur condensation to remove additional sulfur.

Multiple reactors are provided to achieve a more complete conversion of the H_2S.

Condensers are provided after each reactor to condense the sulfur vapor and separate it from the main stream.

    Conversion efficiencies of 94–95% can be attained with two catalytic stages while up to 97% conversion can be attained with three catalytic stages.

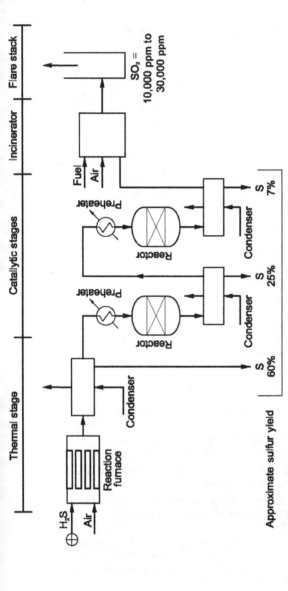

FIGURE 1-10 Simplified process flow schematic for a two-stage Claus process plant.

Dictated by environmental concerns, the effluent gas (SO_2) is either vented, incinerated, or sent to a "tail gas treating unit."

Tail Gas Treating

General Discussion

Many different processes used in tail gas treating today

Processes can be grouped into two categories:

Leading among these processes are the Sulfreen and the Cold Bed Absorption (CBA) processes.

Processes are similar.

Processes utilize two parallel Claus reactors in a cycle, where one reactor operates below the sulfur dew point to absorb the sulfur, while the second is regenerated with heat to recover the absorbed sulfur.

Recoveries up to 99.9% of the inlet sulfur stream are possible.

Second category involves the conversion of the sulfur compounds to H_2S and then the absorbing of the H from the stream.

SCOT process appears to be the leading choice among this type of process.

Uses an amine to remove the H_2S, which is usually recycled back to the Claus plant.

Other types of processes oxidize the sulfur compounds to SO_2 and then convert the SO_2 to a secondary product such as ammonium thiosulfate, a fertilizer.

These plants can remove more than 99.5% of the sulfur and may eliminate the need for incineration.

Costs of achieving tail gas cleanup are high, typically double the cost of a Claus unit.

Sulfa-Check

General Discussion

Converts H_2S to sulfur in a bubble tower of oxidizing and buffering agents

Used in applications with low H_2S concentrations and where regeneration is not required

pH must be held above 7.5

Requires temperatures below 110 °F (42 °C)

Requires pressures above 20 psig (1.35 barg)

Requires 2 lbs of H_2S per gallon of solution at temperatures less than 110 °F (42 °C).

DISTILLATION PROCESS

Ryan-Holmes Distillation Process

General Discussion

Uses cryogenic distillation to remove acid gases from a gas stream

Process is applied to remove CO_2 for LPG separation or where it is desired to produce CO_2 at high pressure for reservoir injection or other use.

Process Description

Process consists of two, three, or four fractionating columns.

Gas stream is first dehydrated and then cooled with refrigeration and/or pressure reduction.

Three-Column System

Used for gas streams containing less than 50% CO_2

First column operates at 450–650 psig (3100–4500 kPa) and separates a high quality methane product in the overhead.

Temperatures in the overhead are from 0 to −140 °F (−18 to −95 °C).

Second column operates at a slightly lower pressure and produces a CO_2 stream overhead, which contains small amounts of H_2S and methane.

Bottom product contains H_2S and the ethane plus components.

Third column produces NGL liquids, which are recycled back to the first two columns.

It is this recycle that allows the process to be successful.

NGL liquids prevent CO_2 solid formation in the first column and aids in the breaking of the ethane/CO_2 azeotrope in the second column to permit high ethane recoveries.

Four-Column System

Used where CO_2 feed concentration exceeds 50%.

The initial column in this scheme is a de-ethanizer.

The overhead product, a CO_2/methane binary, is sent to a bulk CO_2 removal column and de-methanizer combination.

CO_2 is produced as a liquid and is pumped to injection or sales pressure.

Two-Column System

Two-column system is used when a methane product is not required and is thus produced with the CO_2.

Very high propane recoveries may be achieved; however, little ethane recovery is achieved.

These processes require feed gas preparation in the form of compression and dehydration, which add to their cost.

They are finding applications in enhanced oil recovery projects.

GAS PERMEATION PROCESS

Membranes

Definition

Thin semipermeable barriers that selectively separate some compounds from others

Applications

Used primarily for bulk CO_2 removal from natural gas streams

Natural gas upgrading:

Can remove CO_2 and H_2O to pipeline specifications

Lowers H_2S levels

Used onshore/offshore, at wellhead or at gathering facilities.

Enhanced oil recovery (EOR) operations

Used to recover CO_2 from EOR floods for recycle injection into oil reservoir, thus increasing oil recovery

CO_2 is removed from an associated natural gas stream and

Re-injected into an oil to enhance oil recovery.

Materials used for CO_2 removal:

Polymer based (properties modified to enhance performance)

Cellulose acetate (most rugged)

Polyimides, polyamides, polysulfone

Polycarbonates, polyetherimide

Membrane Permeation

Do not act as filters where small molecules are separated from larger ones through a medium of pores

Operate on the principle of solution-diffusion through a nonporous membrane

Highly solubilized components dissolve and diffuse through the membrane.

Relative permeation rates

Most soluble (fastest gases)

H_2O, H_2, H_2S, CO_2, O_2

Least soluble (slowest gases)

N_2, CH_4, C_2+

CO_2 first dissolves into the membrane and then diffuses through it.

Membranes allow selective removal of fast gases from slow gases.

Membranes do not have pores.

Do not separate on the basis of molecular size.

Separation is based on how well different compounds dissolve into the membrane and then diffuse it.

Fick's law (known as Basic Flux Equation)

> Used to approximate the solution-diffusion process.
>
> Expressed as
>
> $$J = (k \times D \times \Delta p)/l \qquad (1\text{-}1)$$
>
> Where
>
>> J = membrane flux of CO_2 (rate/unit area)
>>
>> k = solubility of CO_2 in the membrane
>>
>> D = diffusion coefficient of CO_2 through the membrane
>>
>> Δp = partial pressure difference of CO_2 between the feed (high pressure) and permeate (low pressure) side of the membrane
>>
>> l = membrane thickness

Solubility (k) and diffusion (D) coefficients are usually combined into a new variable called permeability (P).

Fick's law (known as Basic Flux Equation)

Can be split into two portions:

> Membrane dependent portion (P/l), and
>
> Process dependent portion (Δp)

High flux requires:

> Correct membrane material, and
>
> Correct processing conditions.

P/l is not constant and is sensitive to

> Pressure and
>
> Temperature.

Flick's Law equation can be equally written for methane or any other component in the stream.

> Leads to the definition of a second important variable called selectivity (α)

Selectivity (α)

> Ratio of the permeabilities of CO_2 to other components in the stream
>
> Measure of how much better the membrane permeates CO_2 to the compound in question

Example: a CO_2-to-methane selectivity of 30 means that CO_2 permeates the membrane 30 times faster than methane

Important parameters when selecting a membrane:

Permeability (P)

Higher permeability results in

Less membrane area required for a given separation and

Lower system cost.

Selectivity (α)

Higher selectivity results in

Lower losses of hydrocarbons as CO_2 is removed

Higher volume of salable product.

Unfortunately, high CO_2 permeability does not correspond to high selectivity.

Choice must be made between highly elective or permeable membrane or somewhere in-between on both parameters

Unfortunately, high CO_2 permeability does not correspond to high selectivity.

Usual choice is to use a highly selective material and then make it as thin as possible to increase the permeability.

Reduced thickness makes the membrane extremely fragile and therefore unusable.

In the past, membrane systems were not a viable process because the membrane thickness required to provide the mechanical strength was so high that the permeability was minimal.

Asymmetric Membrane Structure

Single polymer consisting of an extremely thin **nonporous layer** mounted on a much thicker and **highly porous layer** of the same material, as opposed to a homogenous structure, where membrane porosity is more-or-less uniform throughout.

Figure 1-11 is an example of an asymmetric membrane

Nonporous layer

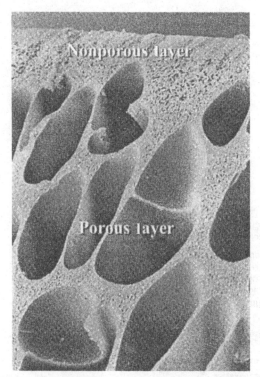

FIGURE 1-11 Asymmetric membrane structure.

Meets the requirements of the ideal membrane:

Highly selective, and

Thin

Porous layer

Provides mechanical support and

Allows the free flow of compounds that permeate through the nonporous layer.

Composite Membrane Structure

Disadvantages of the asymmetric membrane structure

Composed of a single polymer

Expensive to make out of exotic, highly customized polymers and

Produced in small quantities.

Drawback overcome by producing a composite membrane

Consists of a thin selective layer made of one polymer mounted on an asymmetric membrane, which is made of another polymer

Composite structure allows manufacturers to use

Readily available materials for the asymmetric portion of the membrane and

Specially developed polymers, which are highly optimized for the required separation, for the selective layer

Figure 1-12 is an example of a composite membrane structure.

Composite structures are being used in most newer advanced CO_2 removal membranes because the proprieties of the selective layer can be adjusted readily without increasing membrane cost significantly.

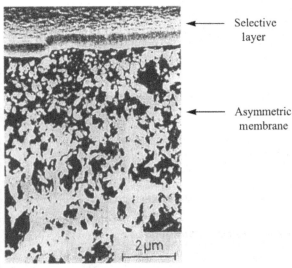

Selective layer

Asymmetric membrane

2 μm

FIGURE 1-12 Composite membrane structure.

Membrane Elements

Manufactured in one of two forms:

Flat sheet

Hollow fiber

Flat Sheet

Combined into a **spiral-wound** element (Figure 1-13).

Two flat sheets of membrane with a permeate spacer in between are glued along three of their sides to form an envelope (or leaf) that is open at one end.

Envelopes are separated by feed spacers and wrapped around a permeate tube with their open ends facing the permeate tube.

Feed enters along the side of the membrane and passes through the feed spacers separating the envelopes.

As the gas travels between the envelopes, CO_2, H_2S, and other highly permeable compounds permeate into the envelope.

Permeated components have only one outlet: they must travel within the envelope to the permeate tube.

Driving force for transport is the low-permeate and high-feed pressures.

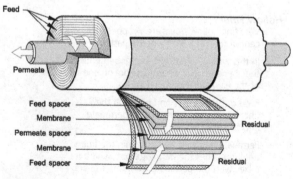

FIGURE 1-13 Spiral wound membrane element.

Permeate gas

Enters the permeate tube through holes drilled in the tube.

Travels down the tube to join the permeate from other tubes.

Any gas on the feed side that does not get a chance to permeate leaves through the side of the element opposite the feed position

Optimization considerations:

Involve the number of envelopes and element diameter

Number of envelopes

Permeate gas must travel the length of each envelope.

Having many shorter envelopes makes more sense than a few longer ones because pressure drop is greatly reduced in the former case.

Element diameter

Larger bundle diameters allow better packing densities but increase the element tube size and decreases cost.

Also increases the element weight, which makes the elements more difficult to handle during installation and replacement.

Hollow Fiber

Very fine hollow fibers are wrapped around a central tube in a highly dense pattern.

In this wrapping pattern, both open ends of the fiber end up at a permeate pot on one side of the element (Figure 1-14).

Feed gas flows over and between the fibers, and some components permeate into them.

Permeate gas then travels within the fibers until it reaches the permeate pot, where it mixes with the permeate from other fibers.

The total permeate exits the element through a permeate pipe.

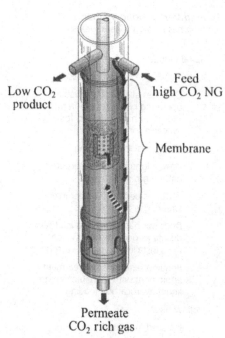

FIGURE 1-14 Hollow-fiber membrane element.

Gas that does not permeate eventually reaches the element's center tube, which is perforated in a way similar to that of the spiral-wound permeate tube.

In this case, however, the central tube is for residual collection, not permeate collection

Optimization considerations

Sleeve design

Forces the feed to flow countercurrent to the permeate instead of the more usual and less efficient cocurrent flow pattern.

Adjusting fiber diameters

Finer fibers give higher packing density but larger fibers have lower permeate pressure drops and so use the pressure driving force more efficiently.

Spiral Wound Versus Hollow Fiber

Each type has its own advantages and limitations.

Spiral wound

Installed in horizontal vessels

Operates at higher allowable operating pressures (75 barg) and thus have higher driving force available for permeation

More resistant to fouling

Have a long history of service in natural gas sweetening

Performs best with colder inlet stream gas temperatures

Does not handle varying inlet feed quality as good as hollow fiber units installed in vertical vessels

Requires extensive pretreatment equipment with high inlet stream liquid hydrocarbon loading

Hollow fiber

Installed in vertical vessels

Offer a higher packing density

Operates at lower inlet stream pressures (40 barg)

Handles higher inlet stream hydrocarbon loading better than spiral wound units

Requires inlet feed gas chilling

Hollow fiber based plants are typically smaller than spiral wound-based plants

Handles varying inlet feed quality better than spiral wound units installed in horizontal vessels

Major Vendors

Spiral wound—UOP, Medal, Kvaerner

Hollow fiber—Cynara (NATCO)

Membrane Modules

Once the membranes have been manufactured into elements, they are joined together and inserted into a tube (Figure 1-15).

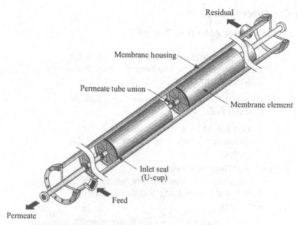

FIGURE 1-15 Membrane module with elements.

Membrane Skids

Multiple tubes are mounted on skids in either a horizontal or vertical orientation, depending on the membrane company.

Figure 1-16 is an example of a skid with horizontal tubes.

FIGURE 1-16 Horizontal membrane skid.

Design Considerations

Process Variables Affecting Design

Flow Rate

Required membrane element area is directly proportional to flow rate since the membrane systems are modular.

Hydrocarbon losses (lost to vent) are directly proportional.

Percentage of hydrocarbon losses (hydrocarbon losses/ feed hydrocarbons) stay the same.

Operating Temperature

An increase in temperature increases membrane permeability and decreases selectivity.

Membrane area requirement is decreased, but the hydrocarbon losses and recycle compressor power for multistage systems are increased (Figure 1-17).

Feed Pressure

An increase in feed pressure

Decreases both membrane permeability and selectivity.

Creates a greater driving force across the membrane that results in a

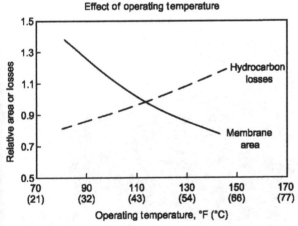

FIGURE 1-17 Effect of operating temperature.

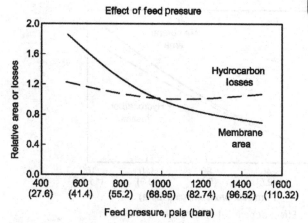

FIGURE 1-18 Effect of feed pressure.

> Net increase in permeation through the membrane results and
>
> Decrease in the membrane area requirements (Figure 1-18).

Increasing the maximum operating pressure results in a cheaper and smaller system.

Limiting factors are the

> Maximum pressure limit for the membrane elements and
>
> Cost and weight of equipment at the higher pressure rating.

Permeate Pressure

Exhibits the opposite effects of feed pressure

Lowering the permeate pressure

> Increases the driving force and
>
> Lowers the membrane area requirements.

Unlike feed pressure, permeate pressure has a strong effect on hydrocarbon losses (Figure 1-19).

> Pressure difference across the membrane is not the only consideration.
>
> Pressure ratio across the membrane is strongly affected by the permeate pressure.

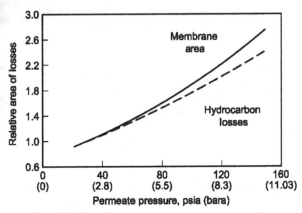

FIGURE 1-19 Effect of permeate pressure.

For example,

> A feed pressure of 90 bar and a permeate pressure of 3 bar produce a pressure ratio of 30.
>
> Decreasing the permeate pressure to 1 bar increases the pressure ratio to 90 and has a dramatic effect on system performance.

Desirable to achieve the lowest possible permeate pressure

> Important consideration when deciding how to further process the permeate stream

For example,

> If permeate stream must be flared, then the flare design must be optimized for low pressure drop.
>
> If permeate stream must be compressed to feed the second membrane stage or injected into a well, the increased compressor horsepower and size at lower permeate pressure must be balanced against the reduced membrane area requirements.

CO_2 Removal

For a constant sales gas CO_2 specification, an increase in feed CO_2

> Increases membrane area requirement and

>> Increases hydrocarbon losses (more CO_2 must permeate, and so more hydrocarbons permeate). This is shown in Figure 1-20.

Membrane area requirement is determined by the percentage of CO_2 removal rather than the feed or sales gas CO_2 specifications themselves.

For example,

> A system for reducing a feed CO_2 content from 10% to 5% is similar in size to one reducing the feed from 50% to 25% or one reducing a feed from 1% to 0.5% if all have a CO_2 removal requirement of about 50%.

For a membrane system, the large difference in percent CO_2 removal (97 versus 70%) means that the system for 0.1% sales gas is about three times the size of the 1% system.

Traditional solvent or absorbent based CO_2 technologies have the opposite limitation.

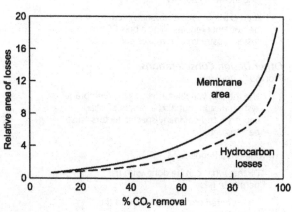

FIGURE 1-20 Effect of CO_2 removal

Their size is driven by the absolute amount of CO_2 that must be removed.

For example,

A system for CO_2 removal from 50% to 25% is substantially larger than one reducing CO_2 from 1% to 0.5%.

For this reason, using membranes for bulk CO_2 removal and using traditional technologies for meeting low CO_2 specifications makes a lot of sense.

Depending on the application, either one or both of the technologies could be used.

Changes in the feed CO_2 content of an existing membrane plant can be handled in a number of ways.

Existing system can be used to produce sales gas with higher CO_2 content.

Additional membrane area can be installed to meet the sales gas CO_2 content, although with increased hydrocarbon losses.

Changes in the feed CO_2 content of an existing membrane plant can be handled in a number of ways.

If heater capacity is available, the membranes can be operated at a higher temperature to also increase capacity.

If an existing nonmembrane system must be de-bottlenecked, installing a bulk CO_2 removal system upstream of it makes good sense.

Other Design Considerations
Process Conditions

Not the only variables affecting the membrane system design, but also a variety of site-, country-, and company-specific factors must be considered.

Environmental Regulations

Dictate what can be done with the permeate gas

Vented (cold or hot vent) to the atmosphere or

Flared either directly or catalytically.

95–99% CO_2 yields low Btu/scf content (flare requires a minimum of 250 Btu/scf to burn).

Location

Dictates a number of issues

Space and weight restrictions

Level of automation

Level of spares that should be available

Single versus multistage operation

Fuel Requirements

Can be obtained

Upstream of the membrane system

Downstream of the pretreatment system

Downstream of the membrane

From the recycle loop in multistage systems

Design Standards

Codes, standards, and recommended practices vary from company to company.

Typical areas that need to be addressed.

Duplex versus carbon steel lines

Maximum pipe velocities

Painting specifications

All items must be predetermined during the bidding stage to prevent costly modifications later.

Process Flow Schemes

Single-Stage Membrane Process (Figure 1-21)
Simplest membrane process.

Feed gas is separated into a permeate stream rich in CO_2 and a hydrocarbon rich residual stream.

Methane loss is approximately 10%.

Multistage Membrane Process
In high CO_2 removal applications

Significant amount of hydrocarbons permeate the membrane and are lost.

Multistage systems attempt to recover a portion of these hydrocarbons.

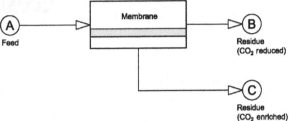

Composition (mole%)	Stream		
	A	B	C
CH$_4$	93.0	98.0	63.4
CO$_2$	7.0	2.0	36.6
Flow rate (MMscfd)	20.0	17.11	2.89
Pressure (psig)	850	835	10

Methane recovery = 90.2%

FIGURE 1-21 Single-stage membrane process.

Two-Step Membrane Process

Allows only a portion of the first stage permeate to be lost

The portion of the first stage permeate that is lost is usually taken from the first membrane modules, where feed CO$_2$, hence permeate CO$_2$, is highest and hydrocarbons lowest.

Rest is recycled to the feed of the first stage.

Permeate that is recycled is at low pressure and must be re-pressurized before it can be combined with the feed gas.

Two-Stage Membrane Process (Figure 1-22)

Processes the first-stage permeate in a second membrane stage.

Permeate from the second stage, which has typically twice the CO$_2$ content as the first stage permeate, is vented.

Multistage Membrane Process

Two-Stage Membrane Process

Residue is either recycled or combined with the feed gas.

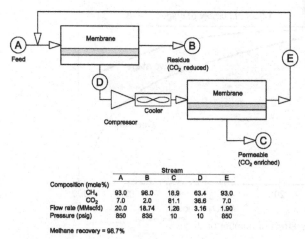

	Stream				
Composition (mole%)	A	B	C	D	E
CH₄	93.0	98.0	18.9	63.4	93.0
CO₂	7.0	2.0	81.1	36.6	7.0
Flow rate (MMscfd)	20.0	18.74	1.26	3.16	1.90
Pressure (psig)	850	835	10	10	850

Methane recovery = 98.7%

FIGURE 1-22 Two-stage membrane process.

Compressor is required to re-pressurize the first stage permeate before it is processed in the second stage.

Two-stage design provides higher hydrocarbon recoveries than two-step or one-stage designs but require more compressor horsepower (because more gas must be compressed to be treated).

Other flow schemes are rarely used.

Two-stage with premembrane flow scheme.

Single-stage provides bulk CO_2 removal, followed by a two-stage system for final CO_2 removal.

Uses a much smaller recycle compressor than that required by a standard two-stage system, although hydrocarbon losses are higher because of the single-stage portion of the system.

Many factors must be considered when deciding whether to use a single-stage or multistage system.

An economic analysis must be done to ensure that the cost of installing and

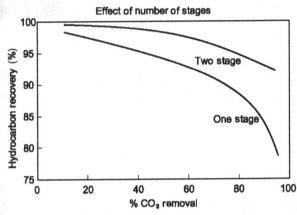

FIGURE 1-23 Effect of number of stages.

operating a recycle compressor does not exceed the savings in hydrocarbon recovery.

Figure 1-23 plots the percentage hydrocarbon recovery versus percentage CO_2 removal for one- and two-stage systems at certain process conditions.

Percentage hydrocarbon recovery is defined as the percentage of hydrocarbon recovered to the sales gas versus the hydrocarbons in the feed gas.

Hydrocarbon recovery of a two-stage is significantly better than that for a single-stage system.

When deciding whether to use a single- or multistage approach, one must also consider the impact of the recycle compressor.

Additional hydrocarbons used as fuel, which increases the overall hydrocarbon losses, as well as the significant capital cost of the compressor and maintenance.

For moderate CO_2 removal applications (less than 50%), single-stage membrane systems usually provide better economic returns than do multistage systems.

Membrane Pretreatment

General Considerations

Proper pretreatment design is critical to the performance of all membrane systems.

Improper pretreatment generally leads to performance decline rather than complete nonperformance.

Substances that lower performance of CO_2 removal

Liquids (water)

Causes swelling and destruction of membrane integrity.

BETEX and heavy hydrocarbons (C_6 to C_{35+})

Forms a film on the membrane surface which drastically decreases permeation rate.

Corrosion inhibitors and well additives

Some are destructive while others are safe.

Consult with manufactures for guidance.

Particulate material

Particles can block the membrane flow area.

Blockage is lower for spiral-wound than for hollow-fiber elements (low flow area).

Long-term particle flow into any membrane could eventually block it.

Pretreatment System Considerations

Must remove above compounds

Must ensure that liquids will not form within the membranes themselves

Two conditions may allow condensation within the membrane:

Gas cools down, as a result of the Joule–Thomson effect, as it passes through the membrane.

Since CO_2 and the lighter hydrocarbons permeate faster than the heavier hydrocarbons, the gas becomes heavier and therefore its dew point increases through the membrane.

Condensation is prevented by achieving a predetermined dew point before the membrane and then heating the gas to provide a sufficient margin of superheat. Heavy hydrocarbon content can vary widely from initial pre-startup estimates and also from month to month during the plant life.

Large variations are seen even between different wells in the same area.

Pretreatment system must account for variations by incorporating a wide safety margin that protects the membranes against a wide range of contaminants.

Traditional Pretreatment

Figure 1-24 shows the equipment used in traditional pretreatment system:

Coalescing filter for liquids and mist elimination

Nonregenerable adsorbent guard bed for trace contaminant removal

Particle filter for dust removal after the adsorbent bed

Heater for providing sufficient superheat to the gas

Figure 1-24 flow scheme is adequate for light, stable composition gases, but it has the following limitations:

Adsorbent bed is the only piece of equipment that is removing heavy hydrocarbons.

A sudden surge in heavy hydrocarbon content or heavier than initially estimated feed gas can saturate the adsorbent bed within days and render it useless.

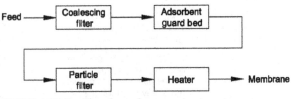

FIGURE 1-24 Traditional membrane pretreatment.

Since the beds are nonregenerable, they can become functional again only after the adsorbent has been replaced.

Problems with the heater require that the whole membrane system be taken offline, because the heater is the only piece of equipment providing superheat.

Additions to Traditional Pretreatment

Additional equipment is added to enhance the system performance.

Chiller

A chiller may be included to reduce the dew point of the gas and the heavy hydrocarbon content.

Since chilling does not completely remove all heavy hydrocarbons, an adsorbent guard bed is still required.

If deep chilling is necessary, steps must be taken to prevent hydrates from forming, either

Dehydrating the gas upstream or

Adding hydrate inhibitors.

If inhibitors are added, they may need to be removed downstream of the chiller because some inhibitors may damage the membrane.

Turbo-Expander

It serves the same purpose as a chiller, but has the benefit of being a dry system.

It is smaller and lighter than refrigeration system.

A disadvantage is the net pressure loss, which must be taken up by the export compressor.

Glycol Unit

It is added upstream of the chiller to prevent hydrate formation or freeze-up.

Adsorbent guard bed are still required to remove heavy hydrocarbons but must be larger than it would normally be because it must also remove glycol carried over from the adsorber vessels.

Enhanced Pretreatment
Need for Enhanced Pretreatment

Not uncommon for initial design basis, based on an extended gas analysis, to differ from actual analysis after the membrane system has been started up.

> For example, feed gas may be heavier than originally anticipated.

Figure 1-25 shows the phase envelopes for the design and actual gas analysis.

Pretreatment system may not have sufficient flexibility to handle a wide departure from design.

> Adsorbent bed may become fully saturated within a short time, leading to performance degradation.

> Preheaters may not be large enough to achieve feed temperatures that are much higher than designed.

> A standard way to handle a gas that is heavier than expected is to operate the membranes at a higher temperature.

> Temperature increase increases the margin between gas dew point and operating temperature and thus prevents condensation in the membrane.

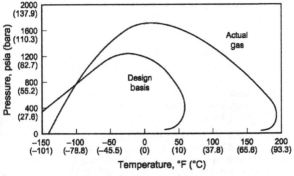

FIGURE 1-25 Expected and actual phase envelopes.

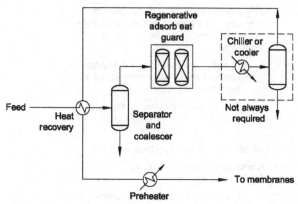

FIGURE 1-26 Enhanced pretreatment flow scheme.

Figure 1-26 shows an enhanced pretreatment scheme that is more suitable for cases where one or more of the following is expected:

Wide variation in feed gas content

Significant amount of heavy hydrocarbons or other contaminants

Feed gas that may be heavier than analyzed, based on the known information from nearby wells or other locations

Feed gas is first cooled down in a heat recovery exchanger, and any condensate formed is removed in a separator and a coalescer.

Liquid-free gas then enters a regenerable adsorbent guard bed system, where heavy hydrocarbons and other harmful components are completely removed.

> Water is removed along with the heavy hydrocarbons, and thus no upstream dehydration is required.

The contaminant-free gas passes through a particle filter leaving the adsorbent guard bed system.

Sometimes the product gas is cooled down in a chiller whose main purpose is to reduce the hydrocarbon dew point of the feed gas.

Any condensate formed in the chiller is removed in a separator and the separator–outlet gas is routed to the feed cross exchanger.

Here the gas cools down the system-feed gas and also obtains necessary superheat.

Further superheat and control of membrane feed temperature are provided by a preheater.

Benefits of enhanced pretreatment are:

Complete removal of heavy hydrocarbons

Unlike other pretreatment schemes, the absolute cutoff of heavy hydrocarbons is possible.

Regenerative system

Since adsorbent guard bed is regenerable, it is better able to handle fluctuations in the heavy hydrocarbon content of the feed gas than traditional guard beds, which require frequent replacement of adsorbent material.

Ability to cope with varying feed composition

Cycle time can be adjusted to provide efficient treatment of a wide variety of feed compositions and heavy hydrocarbon contents.

Reliability

System can be designed to operate satisfactorily even if one of its vessels is taken offline.

Critical items in the pretreatment system are usually spared so they can be serviced or maintained without shutting the system down.

Efficiency

System is able to provide a number of functions, such as removal of water, heavy hydrocarbons, and mercury, that would normally be provided by separate pieces of equipment.

Heat recovery is implemented in the pretreatment scheme as well as within the system itself.

Advantages of Membrane Systems

Membrane systems have major advantages over more traditional methods of CO_2 removal.

Lower Capital Cost (CAPEX)
Membrane systems are skid mounted, except for larger pretreatment vessels.

Scope, cost, and time required for site preparation are minimal.

Installation costs are significantly lower than alternative technologies, especially for remote areas and offshore installations.

Membrane systems do not require the additional facilities, such as solvent storage and water treatment, needed by other processes.

Lower Operating Costs (OPEX)
Only major operating cost for single-stage membrane systems is replacement.

Cost is significantly lower than the solvent replacement and energy costs associated with traditional technologies.

Improvements in membrane and pretreatment design allow a longer useful membrane life, which further reduces operating costs.

Energy costs of multistage systems with large recycle compressors are usually comparable to those for traditional technologies.

Deferred Capital Investment
Gas flow rates often increase over time as more wells are brought online.

With traditional technologies, the system design needs to take this later production into account in the initial design; thus the majority of the equipment is installed before it is even needed.

Modular nature of membrane systems means

Only the membranes that are needed at start-up need be installed.

Rest can be added, either into existing tubes or in new skids, only when they are required.

On offshore platforms, where all space requirements must be accounted for, space can be left for expansion skids rather than having to install them at the start of the project.

High Turndown

Modular nature of membrane systems means that low turndown ratios, to 10% of the design capacity or lower, can be achieved.

Turn up and turndown increments can be set at whatever level is required during the design phase.

Operational Simplicity and High Reliability
Single-Stage Membrane Systems

Single-stage membrane systems have no moving parts.

They have almost no unscheduled downtime.

They are simple to operate.

They can operate unattended for long periods, provided that external upsets, such as well shutdowns, do not occur.

Equipment in pretreatment system that could cause downtime, such as filter coalescers, are usually spared so that production can continue while the equipment is under maintenance.

Addition of a recycle compressor adds some complexity to the system but still much less than with a solvent or adsorbent based technology.

Multistage Membrane Systems

They can be operated at full capacity as single-stage systems when the recycle compressor is down, although hydrocarbon losses will increase

Startup, normal operation, and shutdown of a complex multistage system can be automated so that all important functions are initiated from a control room with minimal staffing

Good Weight and Space Efficiency

Skid construction can be optimized to the space available.

Multiple elements can be inserted into tubes to increase packing density.

Space efficiency is especially important for offshore environments, where deck area is at a premium.

Figure 1-27 illustrates the space efficiency of membrane systems.

The membrane unit in the lower left corner replaced all the amine and glycol plant equipment shown in the rest of the picture.

Adaptability

Since membrane area is dictated by the percentage of CO_2 removal rather than absolute CO_2 removal, small variations in feed CO_2 content hardly change the sales gas CO_2 specification.

For example, a system designed for 10% down to 3% CO_2 removal produces

A 3.5% product from a 12% feed gas and

A 5% product from a 15% feed gas.

Adjusting process parameters such as operating temperature, the designer can further reduce the sales gas CO_2 content.

FIGURE 1-27 Size comparison of membrane and amine systems.

Environmental Friendly

Membrane systems do not involve the periodic removal and handling of spent solvents or adsorbents.

Permeate gases can be

Flared, vented, or reinjected into the well

Used as fuel

Items that do not need disposal, such as spent membrane elements, can be incinerated.

Design Efficiency

Membrane and pretreatment systems integrate a number of operations

Dehydration

CO_2 and H_2S removal

Dew point control

Mercury removal

Traditional CO_2 removal technologies require all of these operations as separate processes and may also require additional dehydration because some technologies saturate the product stream with water.

Power Generation

Permeate gas from membrane systems can be used to provide fuel gas for power generation, either for a recycle compressor or for other equipment.

This virtually free fuel production is especially useful in membrane-amine hybrid systems, where the membrane system provides all the energy needs of the amine system.

Ideal for De-bottlenecking

Since expanding solvent- or adsorbent-based CO_2 removal plants without adding additional trains is difficult, an ideal solution is to use membrane for bulk acid gas removal and leave the existing plant for final cleanup

An additional advantage is that the permeate gas from the membrane system can often be based as fuel for the existing plant, thus avoiding significant increase in hydrocarbon losses.

Ideal for Remote Locations

Many of the factors mentioned above make membrane systems a highly desirable technology for remote locations, where spare parts are rare and labor unskilled.

Solvent storage and trucking, water supply, power generation (unless a multistage system is installed), or extensive infrastructure are not required.

PROCESS SELECTION

Inlet Gas Stream Analysis

An accurate analysis cannot be overstressed.

Process selection and economics depend on knowing all components present in the gas.

Impurities, such as COS, CS_2, and mercaptans (even in small concentrations), can have a significant impact on the design of

> Gas sweetening processes and
>
> Downstream processing facilities.

When sulfur recovery is required, the composition of the acid gas stream feeding the sulfur plant must be considered.

When CO_2 concentrations are greater than 80%

> Selective treating should be considered to raise the H_2S concentration to the sulfur recovery unit (SRU).
>
> It may involve a multistage treating system.

High concentrations of H_2O and hydrocarbons

> Can cause design and operating problems in the SRU.
>
> Effect of these components must be weighed when selecting a gas sweetening process.

Process selection can often be selected based on gas composition and operating conditions.

> High acid gas partial pressures, 345 kPa and above, increase the likelihood of using a physical solvent.
>
> Presence of significant quantities of heavy hydrocarbons in the feed discourages the use of physical solvents.

Low partial pressures of acid gases and low outlet specifications generally require the use of amines for adequate treating.

General Considerations

Each treating processes has advantages relative to the others for certain applications.

When making a final selection, the following facts should be considered:

Type of acid contaminants present in the gas stream

Concentrations of each contaminant and degree of removal required

Volume of gas to be treated and temperature and pressure at which the gas is available

Feasibility of recovering sulfur

Desirability of selectively removing one or more of the contaminants without removing the others

Removal of H_2S to Meet Pipeline Qualities (4 ppm)

Presence and amount of heavy hydrocarbons and aromatics in the gas

Environmental conditions required at the plant site

Feeds with Small Acid Gas Loadings

Batch processes should be considered.

Most common processes include

Iron sponge

Sulfa-Treat

Sulfa-Check

Feeds with Moderate to High Acid Gas Loadings

Disposal and replacement costs are high.

Need to select a process that can be regenerated.

Amine systems are most often used.

DEA is the most commonly used amine.

Acid gas stream off the amine stripper.

Flared at moderate loadings

Converted to elemental sulfur at higher loadings

Process Must Be Added Downstream of the Amine System

Converts acid gas to sulfur

Commonly used systems include

LO-CAT® and

Claus.

Some gas streams can be treated directly with LO-CAT® solution and thus eliminates the need to separate the acid gas components from the gas stream with an amine unit.

When a Claus unit is used, it may be necessary to add tail gas cleanup downstream of the Claus unit if acid gas loadings are very high.

Normally Accomplished with

Amine-Based System

Since the acid gas from the stripper can be vented (assuming levels of H_2S in the gas being treated are very low).

Gas Permeation

Attractive for low volume gas streams in remote areas where the loss of methane is not critical.

Systems with a second stage recycle may be competitive with amine systems.

General Considerations

Often both H_2S and CO_2 are present and must be removed.

Essentially all of the H_2S will have to be removed.

Only a fraction of the CO_2 will have to be removed.

Feeds with Low Concentrations of CO_2

Economical to use a nonselective solvent such as MEA or DEA

Require equipment be sized to essentially remove all the CO_2 so that the H_2S specification can be achieved.

Feeds with Increasing Concentrations of CO_2

Economical to use a selective process such as MDEA, Sulfinol®, Selexol®, etc., which will remove a higher percentage of H_2S than CO_2 from a stream.

Another alternative is to use gas permeation or a carbonate system for bulk removal of CO_2 upstream of a nonselective amine unit.

It may be economical to remove both H_2S and CO_2 to a level where the CO_2 content is acceptable with either a selective or nonselective process, and use a sulfur removal process (iron sponge, Sulfa-Treat, Sulfa-Check, LO-CAT®) for final treating of the residue gas.

Selection Charts

Figures 1-28–1-31 enables one to make a first choice of several potential candidates, which could be investigated to determine which is the most economical for a given set of conditions.

Charts are not meant to replace sound engineering judgment or to cover every possible contingency.

New processes are continuously being developed.

Modifications to existing proprietary products will change their range or applicability and relative cost.

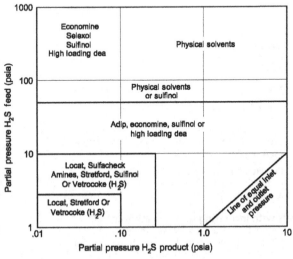

FIGURE 1-28 H_2S removal—no CO_2 present.

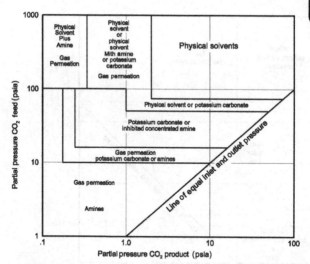

FIGURE 1-29 CO_2 removal—no H_2S.

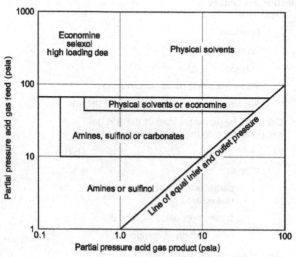

FIGURE 1-30 Removal of H_2S and no CO_2.

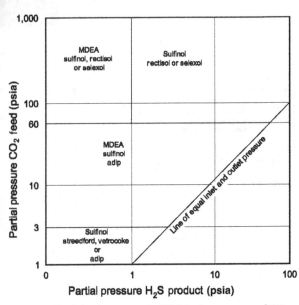

FIGURE 1-31 Selective removal of H_2S in the presence of CO_2.

Selection Procedure
Determine

> Flow rate
>
> Temperature
>
> Pressure
>
> Inlet stream acid gases concentrations
>
> Allowed concentration in the outlet stream

Calculate the partial pressure of the acid gas components using the following equation:

$$PP_i = X_i P \qquad (1\text{-}2)$$

> Where
>
> PP_i = partial pressure of component i, psia (kPa)
>
> P = system pressure, psia (kPa) abs
>
> X_i = mole fraction of component i

Use the appropriate chart, Figures 1-28–1-31, as a guide

DESIGN PROCEDURE

Iron Sponge

General Considerations

Iron sponge process uses a single vessel to contain the hydrated ferric oxide wood shavings.

Inlet gas line should have taps for

Sampling for sulfide

Temperature measurement

Pressure gauge

Injection nozzle for methanol, water or inhibitors

Gas is carried into the top section of the vessel through an inlet nozzle approximately 12 in (0.3 m) above the sponge bed.

Provides uniform flow through the bed, thus minimizing the potential for channeling.

Iron sponge chips are supported by a perforated heavy metal support plate and a coarse support packing material.

This material may consist of scrap pipe thread protectors and 2–3 in (50–75 mm) sections of small diameter pipe.

This provides support for the bed, while offering some protection against detrimental pressure surges.

Gas exits the vessel at the bottom through the vessel side-wall.

This arrangement minimizes entrainment of fines.

Exit line should have a pressure tap and sample test tap.

Access to the vessel is provided through man ways in the top head and bottom head or side of a larger diameter vessel (diameters greater than 36 in (0.92 m) ID).

For small diameter vessels, flanged top and bottom heads offer a viable solution.

Vessel is generally constructed of carbon steel which has been heat treated.

Control of hardness is required because of the potential for sulfide stress cracking.

Vessel is either internally coated, lined, or clad with stainless steel.

Internal coating is normally used for vessels operating below 300 psig (2070 kPa).

Cladding is used for higher operating pressures.

In internally coated units, care must be exercised when setting the coarse packing in order to prevent damage to the coating.

Design Considerations

Determination of the iron sponge vessel diameter

Function of the following variables

Desired bed life

Velocity through the bed

Pressure-drop through the unit

Contact time, and

Channeling potential

Following equations establish the limiting conditions for vessel sizing.

Superficial gas velocity (i.e., gas flow rate divided by vessel cross-sectional area) through the iron sponge bed is normally limited to a maximum of 10 ft/s (3 m/s) at actual flow conditions to promote proper contact with the bed and prevent excessive pressure drop.

Minimum vessel diameter for gas velocity is given by

Oilfield units

$$d_{min} = 60\left(\frac{Q_g TZ}{PV_{g\,max}}\right)^{1/2} \tag{1-3a}$$

SI units

$$d_{min} = 8.58\left(\frac{Q_g TZ}{PV_{g\,max}}\right)^{1/2} \tag{1-3b}$$

Where

d_{min} = minimum internal vessel diameter, cm (in)

Q_g = gas flow rate, std m³/h (MMSCFD)

T = operating temperature, K. (°R)

$$Z = \text{gas compressibility factor}$$
$$P = \text{operating pressure, kPa (psia)}$$
$$V_{gmax} = \text{Maximum gas velocity, m/s (ft/s)}$$

Maximum rate of deposition of 15 grains of $H_2S/min\text{-}ft^2$ (628 grains of H_2S/h m^2) of bed cross-sectional area is also recommended to allow for the dissipation of the heat of reaction.

The following establishes a minimum required diameter for deposition given by

Oilfield units

$$d_{min} = 8945 \left(\frac{Q_g \times H_2S}{\phi} \right)^{1/2} \tag{1-4a}$$

SI units

$$d_{min} = 4255 \left(\frac{Q_g \times H_2S}{\phi} \right)^{1/2} \tag{1-4b}$$

Where

$\phi = $ rate of deposition, grains/h m^2
    (grains/min ft^3)

$H_2S = $ mole fraction of H_2S

The larger of the diameter calculated by Equation (1-3) or (1-4) will set the minimum vessel diameter.

At very low superficial gas velocities less than 2 ft/s (0.61 m/s), channeling of the gas through the bed may occur.

An upper limit to the vessel diameter may be determined by the following equation assuming a minimum velocity of 2 ft/s:

Oilfield units

$$d_{min} = 60 \left(\frac{Q_g TZ}{PV_{gmin}} \right)^{1/2} \tag{1-5a}$$

SI units

$$d_{min} = 8.58 \left(\frac{Q_g TZ}{PV_{gmin}} \right)^{1/2} \tag{1-5b}$$

Where

$d_{max} = $ maximum internal vessel diameter,
    cm (in)

$V_{gmin} = $ minimum gas velocity, m/s (ft/s)

A contact time of 60 sec is considered a minimum in choosing a bed volume.

A larger volume may be considered as it will extend the bed life and thus extend the cycle time between bed changes.

Assuming a minimum contact time of 1 min, any combination of vessel diameter and bed height which satisfies the following is acceptable:

Oilfield units

$$d^2H \geq 3600 \frac{Q_g TZ}{P} \tag{1-6a}$$

SI units

$$d^2H \geq 73.63 \frac{Q_g TZ}{P} \tag{1-6b}$$

Where

d = vesselinternal diameter, cm (in)

H = bedheight, m (ft)

In selecting acceptable combinations, the bed height should be at least 10 ft (3 m) for H_2S removal and 20 ft (6 m) for mercaptan removal.

Produce sufficient pressure drop to assure proper flow distribution over the entire cross-section.

The vessel diameter should be between d_{min} and d_{max}.

Iron sponge is normally sold in the United States by the bushel.

The volume in bushels can be determined from the following equation once the bed dimensions of diameter and height are known:

Oilfield units

$$Bu_m = 7.85 \times 10^{-5} d^2 H \tag{1-7a}$$

SI units

$$Bu = 0.0022 d^2 H \tag{1-7b}$$

Where

Bu = iron spong volume, bushels

Bu_m = bedheight, m (ft)

The amount of iron oxide, which is impregnated on the wood chips is normally specified in units of pounds of iron oxide (Fe_2O_3) per bushel. Common grades are 6.5, 9, 15 or 20 lbs Fe_2O_3/bushel.

Theoretical bed life for the iron sponge between replacements is determined from

Oilfield units

$$t_c = 3.14 \times 10^{-8} \frac{Fe\ d^2 He}{Q_g \times H_2S} \qquad (1\text{-}8a)$$

SI units

$$t_s = 1.48 \times 10^{-6} \frac{Fe\ d^2 He}{Q_g \times H_2S} \qquad (1\text{-}8b)$$

Where

t_c = cycletime, days

Fe = iron sponge content, kg Fe_2O_3/m3 (lbs Fe_2O_3/bushel)

e = efficiency (0.65–0.8)

The Amine System

General Considerations

Equipment and the methods for designing the equipment are similar for both MEA and DEA systems.

For other amine systems, the licensor should be contacted for detailed design information.

Amine Absorbers

Amine absorbers use countercurrent flow through a trayed or packed tower to provide intimate mixing between the amine solution and the sour gas.

Small diameter towers typically use stainless steel packing while larger towers use stainless steel trays.

For a system using the recommended solution concentrations and loadings, a tower with 20 actual trays is normal.

Variations in solution concentrations and loadings may require further investigation to determine the number of trays.

Amine tower diameter sizing is beyond the scope of this section and is best left to the vendor.

Amine absorbers for small gas flow rates commonly include an integral gas scrubber section in the bottom of the tower.

Scrubber will be the same diameter as required for the absorber section.

Gas leaving the scrubber will pass through a mist eliminator and then a chimney tray.

Purpose of this scrubber is to remove entrained water and hydrocarbon liquids from the gas to help protect the amine solution from contamination.

Amine absorbers for larger gas flow rates normally have a separate scrubber vessel or micro-fiber filter separator so that the tower height can be decreased.

This vessel should be designed according to the two-phase separator design guidelines.

For amine systems with large gas flow rates and large amine flow rates, a scrubber should be considered for the outlet sweet gas to recover carryover due to upsets or foaming.

The gas leaving an amine absorber is saturated with water vapor and may require dehydration.

Amine Circulation Rates

The circulation rates for amine systems can be determined from the acid gas flow rates by selecting a solution concentration and an acid gas loading.

The following equations can be used:

Oilfield units

$$L_{MEA} = \frac{112 Q_g X_A}{c \rho A_L} \tag{1-9a}$$

SI units

$$L_{MEA} = \frac{2.55 Q_g X_A}{c \rho A_L} \tag{1-9b}$$

Oilfield units

$$L_{DEA} = \frac{192 Q_g X_A}{c \rho A_L} \tag{1-10a}$$

SI units

$$L_{DEA} = \frac{4.39 Q_g X_A}{c \rho A_L} \tag{1-10b}$$

Where

L_{MEA} = MEA solution circulation rate, m³/h (gpm)

L_{DEA} = DEA solution circulation rate, m³/h (gpm)

Q_g = gas flow rate, std m³/h (MMSCFD)

X_A = required reduction in total acid gas fraction, moles acid gas removed/mole inlet gas. Note: X_A represents moles of all acid components that is, CO_2, H_2S and Meracaptans, as MEA and DEA are not selective

c = amine weight fraction, kg amine/kg solution (lbs amine/ lbs solution)

ρ = solution density, kg/m³ (lbs/gal)

A_L = acid gas loading, mole acid gas/mole amine

The specific gravity of amine solutions at various amine concentrations and temperatures can be found in Figure 1-32 and 1-33.

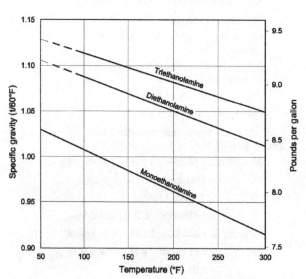

FIGURE 1-32 Specific gravity of amine versus temperature.

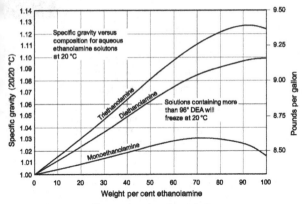

FIGURE 1-33 Specific gravity of amine solution versus composition (courtesy of Jefferson Chemicals)

For design purposes, the following solution strengths and loadings are recommended to provide an effective system without excessive corrosion rates:

MEA solution strength—20 wt% MEA

DEA solution strength—35 wt% DEA

MEA acid gas loading—0.33 mol acid gas/mol MEA

DEA acid gas loading—0.5 mol acid gas/mol DEA

Density of MEA—8.41 lbs/gal

Density of DEA—8.71 lbs/gal

Using the recommended concentrations and specific gravities at 20 °C from Equation 1-33:

$$20\% \text{ MEA} = 1.008 \text{ SG} = 1.008 \times 8.34 \text{ lbs/gal}$$
$$= 8.41 \text{ lbs/gal} = 8.41 \times 0.20$$
$$= 1.68 \text{ lbs MEA/gal}$$
$$= 1.68/61.08 = 0.028 \text{ mol MEA/gal}$$
$$35\% \text{ DEA} = 1.044 \text{ SG} = 1.044 \times 8.34 \text{ lbs/gal}$$
$$= 8.71 \text{ lbs/gal} = 8.71 \text{ lbs/gal} \times 0.35$$
$$= 3.05 \text{ lbs DEA/gal}$$
$$= 3.05/105.14 = 0.029 \text{ mol DEA/gal}$$

Using these design limits, the circulation rates required can be determined from

Equation 1-11 and Equation 1-12:

> Oilfield units
>
> $$L_{MEA} = 202Q_gX_A \qquad (1\text{-}11a)$$
>
> SI units
>
> $$L_{MEA} = 0.038\, Q_gX_A \qquad (1\text{-}11b)$$
>
> Oilfield units
>
> $$L_{DEA} = 126Q_gX_A \qquad (1\text{-}12a)$$
>
> SI units
>
> $$L_{DEA} = 0.024Q_gX_A \qquad (1\text{-}12b)$$

The circulation rate determined with the above equations should be increased by 10–15% to supply an excess of amine.

The rates thus determined can be used to size and select all equipment and piping.

Heat of Reaction

MEA and (MDEA) are basic solutions.

> These solutions react with the hydrogen sulfide and carbon dioxide to form a salt.

The process of absorbing the acid gases generates heat.

The heats of reaction tend to vary with the acid gas loading and the solution concentrations.

With a solution concentration from 15–25 wt% of MEA

> The heat of reaction of H_2S absorbed varies from 550 to 670 Btu/lbs (1,280,000–1,558,000 J/kg) and
>
> The heat of reaction of CO_2 from 620 to 700 Btu/lbs (1,442,000–1,628,000 J/kg).

With a solution concentration of 25–35 wt% of DEA

> The heat of reaction of H_2S absorbed varies from 500 to 600 Btu/lbs (1,163,000–1,396,000 J/kg) and
>
> The heat of reaction of CO_2 from 580 to 650 Btu/lbs (1,349,000–1, 512,000 J/kg).

Table 1-5 Heat of reaction of CO_2 in DEA solutions

Mole Ratio CO_2/DEA	J/kg	Btu/lbs CO_2
35 Weight Percent DEA		
0.2	1,730,000	744
0.4	1,479,000	636
0.5	1,310,000	563
0.6	1,140,000	490
0.8	907,000	390
25 Weight Percent DEA		
0.2	1,593,000	685
0.4	1,384,000	595
0.6	1,103,000	474
0.8	889,000	382

Determine heat of reaction for H_2S absorbed.

Table 1-6 Heat of reaction of H_2S in DEA solutions

Mole Ratio H_2S/DEA	J/kg	Btu/lbs H_2S
0.2	1,405,000	604
0.3	1,342,000	577
0.4	1,279,000	550
0.6	1,177,000	506
0.8	937,000	403
1.0	484,000	208
1.2	368,000	158
1.4	323,000	139

Tables 1-5 and 1-6 give the heat of reaction of CO_2 and H_2S with varying DEA solution concentrations.

The heat of reaction is released when the amine and acid gas first contact and react.

Thus, most of the heating takes place at the bottom of the contactor near the gas entry nozzle.

As the gas goes up the tower it exchanges heat with the amine and leaves the tower at a slightly higher temperature than the inlet amine.

The inlet amine is typically 5.5 °C (10 °F) hotter than the inlet gas.

The amine outlet temperature can be estimated through a heat balance around the column where the heat into the column is the sum of the

Heat in the gas feed inlet

Heat in the amine inlet and

Heat due to the heat of reaction

The heat leaving the column is in the outlet gas stream, the rich amine stream, and column heat losses to the atmosphere.

Flash Vessel

The rich amine solution from the absorber is flashed to remove any absorbed hydrocarbons.

A small amount of acid gases will also flash when the pressure is reduced.

The dissolved hydrocarbons should flash to the vapor phase and be removed.

A small amount of hydrocarbon liquid may begin to collect in this separator.

Thus, provision should be made to remove these liquid hydrocarbons.

Alternately, if the inlet gas to the absorber contains a high percentage of heavier hydrocarbons, a three phase flash vessel may be installed to separate liquid hydrocarbons from the rich amine.

Typically these flash vessels provide 2–3 min of retention time for the amine solution while operating half full.

Determination of the flow rate and composition of the gas flashed in this vessel requires the use of a computer simulation program.

Amine Reboiler

Provides the heat input to an amine stripper, which reverses the chemical reactions and drives off the acid gases.

Heat duty of amine reboilers varies with the system design.

> The higher the reboiler duty, the higher the overhead condenser duty will be.
>
> This gives a higher reflux ratio and thus a smaller column with fewer trays.
>
> The lower the reboiler duty, the lower the reflux ratio will be and thus the tower must have more trays.

Typically for a stripper with twenty trays the reboiler duties will be as follows:

> **MEA system**—1000–1200 Btu/lbs (280–330 MJ/m^3) lean solution.

DEA system—900–1000 Btu/lbs (250–280 MJ/m^3) lean solution.

Thus, the reboiler duty can be estimated:

Equations (1-13) and (1-14)

Oilfield units

$$q_{reb} = 72,000 L_{MEA} \qquad (1\text{-}13a)$$

SI units

$$q_{reb} = 92,905 L_{MEA} \qquad (1\text{-}13b)$$

Oilfield units

$$q_{reb} = 60,000 L_{DEA} \qquad (1\text{-}14a)$$

SI units

$$q_{reb} = 77,421 L_{DEA} \qquad (1\text{-}14b)$$

Where

q_{reb} = reboiler duty, W (Btu/h)

L_{MEA} = MEA circulation rate, m^3/h (gpm)

L_{DEA} = DEA circulation rate, m^3/h (gpm)

Reboilers should be designed to provide the duties shown above.

This will then fix the condenser duty and reflux ratio to match the stripper with twenty trays.

The operating temperature for amine reboilers is determined by the operating pressure and the lean solution concentration.

Typical reboiler temperature ranges are as follows:

MEA reboiler—225–260 °F (107–127 °C)

DEA reboiler—230–250 °F (110–121 °C)

For design purposes, the reboiler temperature for a stripper operating at 10 psig (69 kPa) can be assumed to be 245 °F (118 °C) for 20% MEA, and 250 °F (121 °C) for 35% DEA.

Boiling point versus solution concentration curves at various pressures are shown in Figure 1-34 and 1-35.

Amine Stripper

Use heat from the reboiler to reverse the chemical reactions with CO_2 and H_2S and to generate steam.

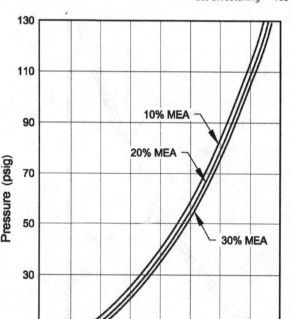

FIGURE 1-34 Boiling points of aqueous monoethanolamine solutions at various pressures.

The steam acts as a stripping gas to remove the CO_2 and H_2S from the liquid solution and to carry these gases to the overhead.

To promote mixing of the solution and the steam, the stripper is a trayed or packed tower, with packing normally used for small diameter columns and trays for larger ones.

Typical stripper operates at 10–15 psig (69–103 kPa) and includes 20 trays, a reboiler, and an overhead condenser.

Rich amine feed is introduced on the third/fourth tray from the top.

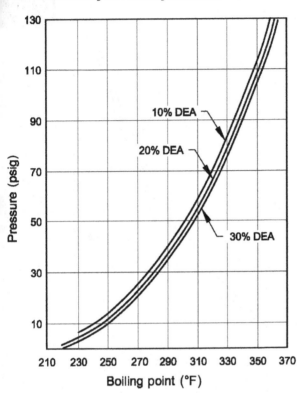

FIGURE 1-35 Boiling points of aqueous diethanolamine solutions at various pressures.

Lean amine is removed at the bottom of the tower and acid gases are removed from the top.

Maximum flow rates within the stripping column can be calculated and then a column size can be determined using the common column sizing methods.

Liquid flow rates are greatest near the bottom tray of the tower where the liquid includes the lean amine flow rate from the tower plus enough water to provide the steam generated by the reboiler.

Lean amine circulation rate is known, and from the reboiler duty, pressure, and temperature, the amount of steam generated can be estimated.

Thus the amount of water can be approximated by assuming all the heat is used to generate steam:

Oilfield units

$$W_{H_2O} = \frac{q_{reb}}{\lambda} \qquad (1\text{-}15a)$$

SI units

$$W_{H_2O} = 3600\frac{q_{reb}}{\lambda} \qquad (1\text{-}15b)$$

Where

W_{H_2O} = water flow rate, kg/h(lbs/h)

q_{reb} = reboiler duty, w (Btu/h)

λ = latent heat of vaporization of water at strippingcolumn pressure, J/kg (Btu/lbs)

The water flow rate in gallons per minute (cubic meters per hour) is approximately

Oilfield units

$$L_{H_2O} = 0.002\frac{q_{reb}}{\lambda} \qquad (1\text{-}16a)$$

SI units

$$L_{H_2O} = 3.6\frac{q_{reb}}{\lambda} \qquad (1\text{-}16b)$$

Where

L_{H_2O} = water flow rate, m^3/h (gpm)

The vapor flow rate within the tower should be calculated at both ends of the column.

The higher of these vapor rates should be used in sizing the tower.

At the bottom of the tower the vapor rate equals the amount of steam generated in the reboiler (Equation 1-15).

Near the top of the tower the vapor rate equals the steam rate overhead plus the acid gas rate.

The steam rate overhead can be approximated from calculating the steam generated in the reboiler (Equation 1-15) and subtracting the amount of steam condensed by raising the amine from its inlet temperature to the reboiler temperature and the amount of steam condensed by vaporizing the acid gases.

Oilfield units

$$W_{steam} = \frac{q_{reb} - (q_{la} - q_{ra} + q_{ag})}{\lambda} \qquad (1\text{-}17a)$$

SI units

$$W_{steam} = 3600\frac{q_{reb} - (q_{la} - q_{ra} + q_{ag})}{\lambda} \qquad (1\text{-}17b)$$

Where

W_{steam} = water rate overhead, kg/h (lbs/h)

q_{reb} = reboiler duty, w (Btu/h)

q_{la} = lean amine solution heat duty, W (Btu/h)

q_{ra} = rich amine solution heat duty, W (Btu/h)

q_{ag} = acid gas heat duty, W (Btu/h)

λ = latent heat of vaporization of water, J/kg (Btu/lbs)

The specific heat of rich amine and lean amine in Btu/lbs °F is shown in Figure 1-36.

Overhead Condenser and Reflux Accumulator

Amine stripper overhead condensers are typically air-cooled fin-fan exchangers.

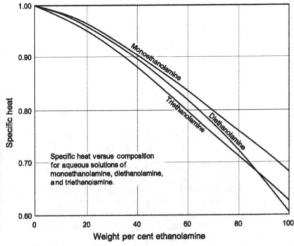

FIGURE 1-36 Specific heat of rich and lean amines.

Once the reboiler duty has been specified, the overhead condenser duty can be determined from a heat balance around the stripper.

A detailed computer simulation can be run to determine the heat balance and the condenser duty.

A simpler method, resulting in a slightly oversized condenser (less than 5%), is to assume that the condenser duty is equal to the regenerator duty minus the sum of the heat required to raise the lean amine from the stripper inlet temperature to the regenerator temperature and the heat of reaction of the acid gases.

Ignores heat in water vapor and acid gases leaving the condenser.

The overhead condenser will cool the vapors leaving the top of the stripper and condense some of the steam for reflux.

The condenser outlet temperature is typically 130–145 °F (54–63 °C) depending upon ambient temperature and is normally designed for an 20–30 °F (11–16 °C) approach to the maximum ambient temperature.

Setting condenser outlet temperature/reflux drum pressure, (0–5 psig) less than the operating pressure of the regenerator, the amount of vapors leaving the condenser can be calculated as follows:

Oilfield units

$$V_R = \frac{(P_R + 14.7)AG}{(P_R + 14.7) - PP_{H_2O}} \times \frac{1}{24} \qquad (1\text{-}18a)$$

SI units

$$V_R = \frac{(P_R + 101.35)AG}{(P_R + 101.35) - PP_{H_2O}} \times \frac{1}{24} \qquad (1\text{-}18b)$$

Where

V_r = mole rate of vapor leaving condenser kg mol/h (lbs mol/h)

P_R = reflux drum pressure, kPa (psig)

A_G = mole acid gas/day, kg mol/day (lbs mol/day)

PP_{H_2O} = partial pressure of water at the condenser outlet temperature, kPa abs (psia)

Amount of reflux can be determined by calculating the amount of steam condensed by the condenser.

Use a top tray temperature of 210 °F (100 °C) as a first assumption, and calculate the amount of heat duty to cool the moles of vapor leaving the condenser from 210 °F (100 °C) to the condenser outlet temperature.

Remaining overhead condenser duty is used in the condensation of steam.

With the use of steam tables the amount of reflux can be calculated as follows:

Oilfield units

$$W_R = 3600 \frac{q_{cond} - q_r}{h_s - h_L} \qquad (1\text{-}19a)$$

SI units

$$W_R = 3600 \left(\frac{q_{cond} - q_r}{h_s - h_L} \right) \qquad (1\text{-}19b)$$

Where

q_{cond} = condenser duty, W (Btu/h)

$\quad q_{vr}$ = heat duty to cool overhead vapors to condenser outlet temperature, W (Btu/h)

$\quad h_s$ = enthalpy of steam at the top tray temperature, W (Btu/lbs)

$\quad h_L$ = enthalpy of water at condenser outlet temperature, W (Btu/lbs)

W_r = reflux rate, kg/h (lbs/h)

The amount of vapor entering the condenser is the sum of the vapor leaving the reflux accumulator and the water condensed for reflux.

From these rates, the partial pressure of water in the vapor leaving can be calculated and the corresponding temperature read from the steam tables for the partial pressure of water.

This is the temperature of the vapor leaving the tower.

If this temperature is different from the one assumed, the calculations should be repeated until the assumed temperature and the calculated temperature are the same.

The reflux accumulator is a separator used to separate the acid gases and steam from the condensed water.

The flow rate of overhead vapor and reflux calculated above can be used to size the reflux accumulator.

A liquid retention time of 3 min should be adequate.

Rich/Lean Amine Exchangers

These exchangers are usually shell and tube exchangers with the corrosive rich amine flowing through the tubes.

The purpose of these exchangers is to reduce the regenerator duty by recovering some of the sensible heat from the lean amine.

The flow rates and inlet temperatures are typically known.

Thus, the outlet temperatures and duty can be determined by assuming an approach temperature for one outlet.

The closer the approach temperature selected, the greater the duty and heat recovered, but the larger and more expensive the exchanger.

An approach temperature of about 30° F (16 °C) provides an economic design, balancing the cost of the rich/lean exchanger and the regenerator to minimize the combined cost of the equipment.

The exchanger duty can be estimated from:

Oilfield units

$$q_{MEA} = 500 L_{MEA} SG_{MEA} C_{PMEA} \Delta T \qquad (1\text{-}20a)$$

SI units

$$q_{MEA} = 0.277 L_{MEA} SG_{MEA} C_{PMEA} \Delta T \qquad (1\text{-}20b)$$

Oilfield units

$$q_{DEA} = 500 L_{DEA} SG_{DEA} C_{PDEA} \Delta T \qquad (1\text{-}21a)$$

SI units

$$q_{DEA} = 0.277 L_{DEA} SG_{DEA} C_{PDEA} \Delta T \qquad (1\text{-}21b)$$

Where

q_{MEA} = MEA exchanger duty, W (Btu/h)

q_{DEA} = DEA exchanger, W (Btu/h)

$$L_{MEA} = \text{MEA circulation rate, m}^3\text{/h (gpm)}$$

$$L_{DEA} = \text{DEA circulation rate, m}^3\text{/h (gpm)}$$

$$C_{PMEA} = \text{MEA specific heat, kJ/kg K (Btu/lbs°F)}$$

$$C_{PDEA} = \text{DEA specific heat, kJ/kg K (Btu/lbs°F)}$$

$$\Delta T = \text{amine temperature change, °C (°F)}$$

$$SG_{MEA} = \text{specific gravity of MEA}$$

$$SG_{DEA} = \text{specific gravity of DEA}$$

Amine Cooler

Proper control of the lean amine temperature entering the absorber is essential for the efficient operation of any amine plant.

The amine cooler is typically an air cooler which lowers the lean amine temperature before it enters the absorber.

The lean amine entering the absorber should be approximately 10 °F (5 °C) warmer than the sour gas entering the absorber.

Lower amine temperatures cause the gas to cool in the absorber and thus condense hydrocarbons, which tends to cause foaming.

Higher temperatures raise the amine vapor pressure, resulting in increased amine losses.

Higher amine temperatures will also increase the outlet gas temperature and increase the water content of the gas, increasing loading of downstream dehydration equipment.

The duty for the cooler can be calculated from the lean amine flow rate, the lean amine temperature leaving the rich/lean exchanger, and the sour gas inlet temperature as follows:

Oilfield units

$$q_{cooler} = 500 L_{LA} SG C_{PLA}(T_{out} - T_m) \qquad (1\text{-}22a)$$

SI units

$$q_{cooler} = \frac{L_{LA} SG C_{PLA}(T_{out} - T_{in})}{3600} \qquad (1\text{-}22b)$$

Where

$$q_{cooler} = \text{lean amine cooler duty, W (Btu/h)}$$

$$SG = \text{specific gravity of lean amine}$$
$$\text{(water} = 1.0)$$

L_{LA} = lean amine circulation rate, m^3/h (gpm)

C_{PLA} = specific heat of lean amine, kJ/kg K (Btu/lbs)

T_{out} = amine cooler outlet temperature = feed gas temperature + 2.2 °C (°F)

T_{in} = amine cooler outlet temperature = temperature out of rich/lean amine exchanger, °C (°F)

Amine Solution Purification

Due to side reactions and/or degradation a variety of contaminants will accumulate in an amine system.

The method of removing these depends on the amine involved.

In an MEA process, when the contaminants COS and CS_2 are present in the acid gas stream, an undesirable side reaction occurs, resulting in the formation of heat stable salts.

These salts should be removed from the system. For this reason, MEA systems usually include a reclaimer.

The reclaimer is a kettle type reboiler operating on a small side stream of lean solution.

The temperature in the reclaimer is maintained such that the water and MEA boil to the overhead and back to the stripper, leaving the heat stable salts in the reclaimer

Once full, the reclaimer is shut in and dumped to a waste disposal. The impurities are removed from the system, but the MEA bonded to the salts is also lost

For DEA systems, a reclaimer is not required because the reactions with COS and CS_2 are reversed in the stripper. The small amount of degradation products from CO_2 can be removed by a carbon filter on a side stream of lean solution.

Amine Solution Pumps

A large portion of an amine plant's energy demand is created by the amine solution pumps.

Usually, a booster pump and main circulation pump are included in the amine process.

The booster pump, located downstream of the amine stripper, provides enough head, typically 25–40 psi (72–275 kPa), to flow the lean amine through filtration equipment, lean amine/rich amine exchanger, the amine cooler and into the lean amine surge tank.

Booster pumps are usually centrifugal type, and it is common to provide 100% spare capacity due to the service conditions. The booster pump horsepower can be estimated as follows:

Oilfield units

$$BHP = \frac{\Delta P L_{LA}}{1714e} \qquad (1\text{-}23a)$$

SI units

$$BHP = \frac{\Delta P L_{LA}}{3598e} \qquad (1\text{-}23b)$$

Where

BHP = pump brake horsepower, kW (HP)

ΔP = differential pressure, kPa (psi)

L_{LA} = lean amine circulation rate, m³/h (gpm)

e = pump efficiency

: 0.7 for centrifugal pumps

: 0.9 for reciprocating pumps

The amine circulation pumps take suction from the lean amine surge tank and boost the amine to the operating pressure of the absorber.

Both centrifugal and positive displacement pumps are used in this application.

The type selected usually depends on absorber operating pressure.

A typical arrangement calls for three 50% capacity pumps to provide spare capacity.

The circulation pump horsepower may be estimated using the same formula as for booster pumps, but with the ΔP required to raise pressure from the surge tank operating pressure to the contactor operating pressure.

Procedure for Sizing an Amine System

Determine a flash tank size.

Determine rich/lean amine exchanger duty.

Set the stripper overhead condenser outlet temperature based on a 20–30 °F (11–17 °C) approach to the maximum ambient temperature and calculate the flow rates of acid gases and steam.

Determine the condenser duty and reflux rate by performing an energy balance around the stripper.

Size the reflux accumulator.

Determine the lean amine cooler duty and the lean amine temperature leaving the rich/lean amine exchanger determined earlier.

Determine horsepower requirements for the booster pump and main circulation pump using Equation 1-23.

DESIGN EXAMPLES (OILFIELD UNITS)

Example Problem 1: Iron Sponge Unit

Given:

Q_g = 2 MMSCFD

SG = 0.6

H_2S = 19 ppm

P = 1200 psig

T = 100 °F

Mercaptans are not present.

Solution

Step 1. Calculate Minimum Vessel Diameter For Gas Velocity (Equation 1-3a)

Oilfield units

$$d_{min} = 60\left(\frac{Q_g TZ}{PV_{g\,max}}\right)^{1/2}$$

Where

d_{min} = minimum internal vessel diameter, in

Q_g = gas flow rate, MMSCFD

T = operating temperature, °R

Z = gas compressibility factor

(*Source:* Surface Production Operations, Volume 1, 3rd edition, Figure 3-9)

$P =$ Operating pressure, psia

$V_{gmax} =$ maximum gas velocity, ft/s

Oilfield units

$$d_{min} = 60\left(\frac{Q_g TZ}{PV_{g\,max}}\right)^{1/2}$$

$$Z = 0.85$$

Use $V_{gmax} = 10\ ft/s$

$$d_{min} = 60\left(\frac{(2)(100+460)(0.85)}{(1200+14.7)(10)}\right)^{1/2}$$
$$d_{min} = 16.8\ in$$

Step 2. Calculate Minimum Vessel Diameter For Deposition (Equation 1-4a)

Oilfield units

$$d_{min} = 8945\left(\frac{Q_g X_{H_2S}}{\phi}\right)^{1/2}$$

Where

$\phi =$ rate of deposition grains/min ft^2

$X_{H_2S} =$ mole fraction of H_2S

$$d_{min} = 8945\left(\frac{Q_g X_{H_2S}}{\phi}\right)^{1/2}$$

Use a rate of deposition, of 15 grains/min-ft^2

$$d_{min} = 8945\left(\frac{(2)(0.000019)}{15}\right)^{1/2}$$
$$d_{min} = 14.2\ in$$

Step 3. Calculate Maximum Diameter (Equation 1-5a)

Oilfield units

$$d_{max} = 60\left(\frac{Q_g TZ}{PV_{g\,min}}\right)^{1/2}$$

Where

$d_{max} =$ maximum internal vessel diameter, in

$V_{gmin} =$ minimum gas velocity, ft/s $= 2\ ft/s$

$$d_{min} = 60 \left(\frac{(2)(100 + 460)(0.85)}{(1200 + 14.7)(2)} \right)^{1/2}$$

$$= 37.6 \text{ in}$$

Therefore, any diameter from 16.8 to 37.6 in. is acceptable.

Step 4. Choose A Cycle Time of One Month or Longer (Equation 1-6a)

Oilfield units

$$t_c = 3.14 \times 10^{-8} \frac{Fed^2 \, He}{Q_g X_{H_2S}}$$

Where

t_c = cycle time, days

Fe = iron sponge content, lbs Fe_2O_3/bushel

e = efficiency (0.65 tp 0.8)

$$d^2H = \left(\frac{t_c Q_g X_{H_2S}}{3.14 \times 10^{-8} \, Fee} \right)$$

$$d^2H = \frac{(30)(2)(0.000019)}{(3.14 \times 10^{-8})(9)(0.65)}$$

$$d^2H = 6206$$

d (in.)	H (ft)
18	19.15
20	15.52
22	12.82
24	10.77
30	6.90
36	4.79

An acceptable choice is a 30 in. O.D. vessel. Since t_c and e are arbitrary, a 10-ft bed is appropriate.

Step 5. Calculate Volume of Iron Sponge to Purchase (Equation 1-7a)

Oilfield units

$$Bu = 0.0044d^2H$$

Where

Bu = iron sponge volume, bushels

$$Bu = 0.0044 \, d^2H$$

$$Bu = 0.0044(30)^2 (10)$$

$$Bu = 39.6 \text{ bushels}$$

Example Problem 2: Amine Processing Unit (DEA)

Given:

Gas Volume	= 100 MMSCFD
Gas Gravity	= 0.67 SG (Air = 1.0)
Pressure	= 1000 psig
Gas Temperature	= 100 °F
CO_2 inlet	= 4.03%
CO_2 outlet	= 2%
H_2S inlet = 19 ppm	= 0.0019%
H_2S outlet	= 4 ppm
Max. Ambient Temp.	= 100 °F

Solution:

Step 1. Process Selection

Total acid gas inlet = 4.03 + 0.0019 = 4.032%

Partial pressure of inlet acid gas = 1015 × (4.032/100) = 40.9 psia

Total acid gas outlet = 2.0%

Partial pressure of outlet acid gas = 1015 × (2.0/100) = 20.3 psia

From Figure 1-29 for removing CO_2 and H_2S, possible processes are: Amines, Sulfinol®, and Carbonates.

Step 2. DEA Circulation Rate

Determine the circulation rate
Equation 1-10a:

Oilfield units

$$L_{DEA} = \frac{192 Q_g X_A}{c \rho A_L}$$

Where

L_{DEA} = DEA solution circulation rate, m^3/h (gpm)

Q_g = gasflow rate, std m³/h (MMSCFD)

X_A = required reduction in total acid gas fraction, moles acid gas removed/mole inlet gas. Note: X_A represents moles of all acid components, that is, CO_2, H_2S and Meracaptans, as MEA and DEA are not selective

c = amine weight fraction, kg amine/kg solution (lbs amine/ lbs solution)

ρ = solution density, kg/m³ (lbs/gal)

A_L = acid gas loading, mole acid gas/ mole amine

r = DEA density

   = 8.71 lbs/gal

c = 0.35 lbs/lbs

A_L = 0.50 mol/mol

Q_g = 100 MMSCFD

X_A = 4.032% = 0.04032

Note: In order to meet the H_2S outlet, virtually all the CO_2 must be removed, as DEA is not selective for H_2S.

$$L_{DEA} = \frac{192(100)(0.04032)}{(0.35)(8.71)(0.50)} = 508 \text{ gpm}$$

Add 10% for safety = 560 gpm

Step 3. Reboiler Duty

Determine the reboiler duty, Equation 1-14a:

Oilfield units

$q_{reb} = 60{,}000\ L_{DEA}$

Where

q_{reb} = reboiler duty, Btu/h

L_{DEA} = DEA circulation rate (gpm)

$q_{reb} = 60{,}000(560)$
$q_{reb} = 33.6$ MMBtu/h

Step 4. Absorber Heat Balance

Perform a heat balance around the absorber as follows:

Set lean amine inlet temperature at 10 °F higher than inlet gas, or 110 °F.

Assume gas leaving absorber has 5 °F approach to inlet amine, or 115 °F exit temperature.

Determine heat of reaction for CO_2 absorbed using loading and circulation with safety factor considered.

$$\frac{\text{Moles } CO_2}{\text{absorbed}} = \frac{100 \text{ MMSCFD} \times 0.04032}{379.5 \text{ SCF/mol}}$$

$$= 10,620 \text{ mol/day } CO_2$$

moles DEA circulated = 560 gpm × 8.71 lbs/gal × 0.35 lbs/lbs × mol/105.14 lbs × 1440 min/day = 23,400 mol DEA/day

Loading moles CO_2/mol DEA = 10,620/23,400 = 0.45

From Table 1-5, using 35 wt% DEA, the heat of reaction for CO_2 is 592 Btu/lbs

Heat of reaction due to the CO_2 is 592 Btu/lbs × 10,620 mol/day × 1/24 × 44 lbs/mol = 11.5 MMBtu/h

$$\frac{\text{Moles } H_2S}{\text{absorbed}} = \frac{100 \text{ MMSCFD} \times (0.000019 - 0.000004)}{379.5 \text{ SCF/mol}}$$

$$= 3.95 \text{ mol/day } H_2S$$

600 Btu/lbs H_2S × 3.95 × 1/24 × 34 lbs/mol = 3,360 Btu/h

This is insignificant and can be ignored.

Calculate the DEA outlet temperature at the contactor.

Moles of air per SCF = 28.96

$$\text{Gas flow rate} = \frac{100 \text{ MMSCFD} \times 0.67 \times 28.96}{24 \times 379.5}$$

$$= 213,000 \text{ lbs/h}$$

Heat gained by gas steam

$Q = 213,000$ lbs/h × (115 − 100)°F
     × 0.65 Btu/lbs °F
   = 2.08 MMBtu/h

Heat lost to atmosphere—this depends on atmospheric temperature, surface area, wind velocity, etc. Assume 5% of reaction heat lost to atmosphere for an un-insulated absorber.

Total heat gained in outlet amine = Heat of reaction − heat gained by gas stream − heat lost to atmosphere.

$$\text{Heat gained} = 11.5 \text{ MMBtu/h} - 2.08 \\ - (11.5 \times 0.05) \\ = 8.9 \text{ MMBtu/h}$$

$$\begin{array}{l}\text{Rich amine} \\ \text{outlet} \\ \text{temperature}\end{array} = \dfrac{8.9 \text{MMBtu/h}}{\begin{array}{l}560 \text{gpm} \times \\ 8.71 \text{lbs/gal} \times \\ 60 \text{min/h} \times \\ 915 \text{Btu/lbs}°F\end{array}} + 110°F = 143°F$$

Step 5. Flash Tank

Determine flash tank size

Operating pressure = 150 psig

Operating temperature = 143 °F

Amine rate = 560 gpm

$$\begin{array}{l}\text{(Max)}CO_2 \\ \text{flashed}\end{array} = \dfrac{10,620 \text{ mol/day}}{1440 \text{ min/day}} \times 44 \text{ lbs/mol} \\ = 325 \text{ lbs/min}$$

(Max)H_2S absorbed = negligible

Size for 3 min retention time operating 1/2 full.

Step 6. Rich/Lean Exchanger

Determine the rich/lean amine exchanger duty:

$$\begin{array}{l}\text{Lean amine} \\ \text{flow}\end{array} = 560 \text{ gpm} \times 8.71 \times 60 \\ = 293,000 \text{ lbs/h}$$

$$\begin{array}{l}\text{Rich amine} \\ \text{flow}\end{array} = 293,000 + \dfrac{\begin{array}{l}10,620 \text{ mol} \\ CO_2/\text{day} \times \\ 44 \text{ lbs/mol}\end{array}}{24 \text{ h/day}} + \\ \dfrac{4 \text{ mol } H_2S/\text{day}}{24 \text{ h/day}} \times 34 \text{ lbs/mol} \\ = 312,000 \text{ lbs/h}$$

Assume DEA reboiler temperature is 250 and 30 °F approach of rich amine to lean amine.

Rich amine inlet temp = 143 °F (from contactor)

Rich amine outlet temp = 250 − 30 °F

$$= 220 \, °F$$

Calculate lean amine outlet temperature, assuming specific heat of rich amine is the same as the specific heat of lean amine = 0.915 Btu/lbs °F

$$T_{out} = 250 \, °F - \left[(220 - 143) \times \frac{312,000}{293,000} \right]$$

$$= 168 \, °F$$

Exchanger duty = 312,000 lbs/h × 0.915 × (220 − 143)

$$= 22 \, MMBtu/h$$

Step 7. Stripper Overhead

Set the stripper overhead condenser outlet temperature and calculate the flow rates of acid gas and steam. Size the condenser and determine the reflux rate. The acid gas and steam will be vented, flared, or processed further for removal of H_2S.

Set condenser temperature at 30 °F above maximum atmospheric temperature, that is, 130 °F.

From steam tables, the partial pressure of water at 130 °F is 2.22 psia.

Stripper and reflux drum will operate at 10 psig (24.7 psia).

Calculate the vapor rate leaving the reflux condenser:

Oilfield units

$$V_R = \frac{(PR + 14.7)AG}{(PR + 14.7) - PP_{H_2O}} \times \frac{1}{24}$$

Where

V_r = mole rate of vapor leaving condenser, lbs mol/h

P_R = reflux drum pressure, psig

A_G = moles acid gas/day, lbs mol/day

PP_{H_2O} = partial pressure of water at the condenser outlet temperature, psia

$$V_R = \frac{24.7 \times (\text{mol } CO_2 + \text{mol } H_2S)}{24.7 - PP_{H_2O}}$$

$$V_R = \frac{24.7 \times (10,620 + 4)}{24.7 - 2.22}$$
$$= 11,700 \text{ mol/day}$$

$$V_{steam} = 11,700 - 10,624 = 1100 \text{ mol/day}$$

$$\begin{aligned} W_{H_2O} &= 1100 \text{ mol/day} \times 18 \text{ lbs/mol} \\ &= 19,800 \text{ lbs/day} \\ &= 800 \text{ lbs/h lost from condensor} \end{aligned}$$

Step 8. Condenser Duty and Reflux Rate

Determine the condenser duty and reflux rate by performing an energy balance around the stripper.

$$q_{reb} = q_{steam} + q_{H_2S} + q_{CO_2} + q_{amine} + q_{cond}$$

The heat required to vaporize the acid gases (reverse the reaction) is

$$q_{CO_2} = 11.5 \text{MMBtu/h, and } q_{H_2S} = 3 \text{Mbtu/h}$$

Calculate the heat differential between the heat in and the heat out for the lean DEA

$$\begin{aligned} q_{la} &= 293,000 \text{ lbs/h} \times 0.915(250 - 220) \\ &= 8 \text{ MMBtu/h} \end{aligned}$$

$$q_{cond} = 33.6 - 8 - 11.5 \text{ MBtu/h}$$

$$q_{cond} = 14 \text{ MMBtu/h}$$

The reflux condenser must cool the acid gas and steam from the top tray temperature to 130 °F and condense the amount required for reflux.

Assume a top tray temperature of 210 °F.

Heat required to cool the acid gas is approximately

$$q_{ag} = \frac{10,624 \text{ mol/day}}{24 \text{ h/day}} \times 44$$

$$\times 0.65 \text{ Btu/lbs °F} \times (210 - 130)$$
$$= 1.01 \text{ MMBtu/h}$$

$$W_{reflux} = \frac{(14 - 1.01) \text{ MMBtu/h}}{(1149.7 - 180) \text{ Btu/lbs}} = 13,324 \text{ lbs/h}$$

Calculate the vapors leaving the top tray

$$V_{top} = 10{,}624 \text{ mol/day AG} + \frac{(13{,}368 + 788)}{18 \text{ lbs/mol}} \times 24$$

$$= 29{,}440 \text{ mol/day vapors steam and acid gas}$$

$$\text{Moles/day water} = 29{,}440 - 10{,}624$$
$$= 18{,}816$$

$$PP_{H_2O} \text{ in top vapors} = \frac{(18{,}816) \times 24.7 \text{ psia}}{29{,}440 \text{ mol/day total}}$$

$$= 15.8 \text{ psia}$$

This PP_{H_2O} of 15.8 psia is equivalent to a temperature of about 214 °F from Table 1-7.

Recalculate, assuming top tray temperature of 214 °F.

Then heat required to cool the acid gas is:

$$q_{ag} = \frac{10{,}624}{24} \times 44 \times 0.65 \times (214 - 130)$$
$$= 1.060 \text{ MMBtu/h}$$

$$W_{reflux} = \frac{(14 - 1.06) \text{ MMBtu/h}}{1151.2 - 180} = 12{,}290 \text{ lbs/h}$$

$$V_{top} = 10{,}624 \text{ mol/}D_{AG} + \frac{(12{,}244 + 788)}{78} \times 24$$
$$= 28{,}100 \text{ mol/day}$$

$$\text{mol/day water} = 28{,}100 - 10{,}624 = 17{,}480$$

$$PP_{H_2O} \text{ in top vapors} = \frac{17{,}480}{28{,}100} \times 24.7$$
$$= 15.4 \text{ psia}$$

This is equivalent to a top tray temperature of 214.3 °F.

Step 9. Reflux Accumulator

Size the reflux accumulator using the principles of two phase separators for the following conditions:

Liquid volume = 12,290 lbs/h water

Table 1-7 Properties of dry saturated steam

Temp. (°F) t	Abs. Pres. (psia) P	SPECIFIC VOLUME (CU FT/LBS)			ENTHALPY (BTU/LBS)			ENTROPY (BTU/LBS)			Temp. (°F) t
		Sat. Liquid V_f	Evap. V_{fg}	Sat vapor V_g	Sat. Liquid h_f	Evap. h_{fg}	Sat. Vapor h_g	Sat. Liquid s_f	Evap. s_{fg}	Sat. Vapor s_g	
32	0.08854	0.01602	3306	3306	0.00	1075.8	1075.8	0.0000	2.1877	2.1877	32
35	0.09995	0.01602	2947	2947	3.02	1074.1	1077.1	0.0061	2.1709	2.1770	38
40	0.12170	0.01602	2444	2444	8.05	1071.3	1079.3	0.0162	2.1436	2.1597	40
45	0.14752	0.01602	2036.4	2036.4	13.06	1068.4	1081.5	0.0262	2.1167	2.1429	45
50	0.17811	0.01603	1703.2	1703.2	18.07	1065.6	1083.7	0.0361	2.0903	2.1264	50
60	0.2563	0.01604	1206.6	1206.7	28.06	1059.9	1088.0	0.0555	2.0393	2.0948	80
70	0.3631	0.01606	867.8	867.9	38.04	1054.3	1092.3	0.0745	1.9902	2.0647	70
80	0.5069	0.01608	633.1	633.1	48.02	1048.6	1096.6	0.0932	1.9428	2.0360	80
90	0.6982	0.01610	468.0	468.0	57.99	1042.9	1100.9	0.1115	1.8972	2.0087	90
100	0.9492	0.01613	350.3	350.4	67.97	1037.2	1105.2	0.1295	1.8531	1.9826	100
110	1.2748	0.01617	265.3	265.4	77.94	1031.6	1109.5	0.1471	1.8106	1.9577	110
120	1.6924	0.01620	203.25	203.27	87.92	1025.8	1113.7	0.1645	1.7694	1.9339	120
130	2.2225	0.01625	157.32	157.34	97.90	1020.0	1117.9	0.1816	1.7296	1.9112	130
140	2.3886	0.01629	122.99	123.01	107.89	1014.1	1122.0	0.1984	1.6910	1.8894	140
150	3.718	0.01634	97.06	97.07	117.89	1008.2	1126.1	0.2149	1.6537	1.8685	150
160	4.741	0.01639	77.27	77.29	127.89	1002.3	1130.2	0.2311	1.6174	1.8485	160
170	5.992	0.01645	62.04	62.06	137.90	996.3	1134.2	0.2472	1.5822	1.8293	170
180	7.510	0.01651	50.21	50.23	147.92	990.2	1138.1	0.2630	1.5480	1.8109	180
190	9.339	0.01657	40.94	40.96	157.95	984.1	1142.0	0.2785	1.5147	1.7932	190

(Continued)

Table 1-7 Properties of dry saturated steam—cont'd

Temp. (°F) t	Abs. Pres. (psia) P	SPECIFIC VOLUME (CU FT/LBS)			ENTHALPY (BTU/LBS)			ENTROPY (BTU/LBS)			Temp. (°F) t
		Sat. Liquid V_f	Evap. V_{fg}	Sat vapor V_g	Sat. Liquid h_f	Evap. h_{fg}	Sat. Vapor h_g	Sat. Liquid s_f	Evap. s_{fg}	Sat. Vapor s_g	
200	11.526	0.01663	33.62	33.64	167.99	977.9	1145.9	0.2938	1.4824	1.7762	200
210	14.123	0.01670	27.80	27.82	178.05	971.6	1149.7	0.3090	1.4508	1.7598	210
212	14.696	0.01672	26.78	26.80	180.07	970.3	1150.4	0.3120	1.4446	1.7566	212
220	17.186	0.01677	23.13	23.15	188.13	965.2	1153.4	0.3239	1.4201	1.7440	220
230	20.780	0.01684	19.365	19.382	198.23	958.8	1157.0	0.3387	1.3901	1.7288	230
240	24.969	0.01692	16.306	16.323	208.34	952.2	1160.5	0.3531	1.3609	1.7140	240
250	29.825	0.01700	13.804	13.821	216.48	945.5	1164.0	0.3675	1.3323	1.6998	250
260	35.429	0.01709	11.746	11.763	228.64	938.7	1167.3	0.3817	1.3043	1.6860	260
270	41.858	0.01717	10.044	10.061	238.84	931.8	1170.6	0.3958	1.2769	1.6727	270
280	49.203	0.01726	8.628	8.645	249.06	924.7	1173.8	0.4096	1.2501	1.6597	280
290	57.556	0.01735	7.444	7.461	259.31	917.5	1176.8	0.4234	1.2238	1.6472	290
300	67.013	0.01745	6.449	6.466	269.59	910.1	1179.7	0.4369	1.1980	1.6350	300
310	77.68	0.01755	5.609	5.625	279.92	902.6	1182.5	0.4504	1.1727	1.6231	310
320	89.66	0.01765	4.896	4.914	290.28	894.9	1185.2	0.4637	1.1478	1.6115	320
330	103.06	0.01776	4.269	4.307	300.68	887.0	1187.7	0.4769	1.1233	1.6002	330
340	118.01	0.01787	3.770	3.788	311.13	879.0	1190.1	0.4900	1.0992	1.5891	340
350	134.63	0.01799	3.324	3.342	321.63	870.7	1192.3	0.5029	1.0754	1.5783	350
360	153.04	0.01811	2.939	2.957	332.18	862.2	1194.4	0.5158	1.0519	1.5677	360
370	173.37	0.01823	2.606	2.625	342.79	853.5	1196.3	0.5286	1.0287	1.5573	370
380	195.77	0.01836	2.317	2.335	353.45	844.6	1198.1	0.5413	1.0059	1.5471	380
390	220.37	0.01850	2.0651	2.0836	364.17	835.4	1199.6	0.5539	0.9832	1.5371	390

400	247.31	0.01864	1.8447	1.8633	374.97	826.0	1201.0	0.5664	0.9608	1.5272	400
410	276.75	0.01878	1.6512	1.6700	385.83	816.3	1202.1	0.5788	0.9386	1.5174	410
420	308.83	0.01894	1.4811	1.5000	396.77	806.5	1203.1	0.5912	0.9166	1.5078	420
430	343.72	0.01910	1.3308	1.3499	407.79	796.0	1203.8	0.6035	0.8947	1.4982	430
440	381.59	0.01926	1.1979	1.2171	418.90	785.4	1204.3	0.6158	0.8730	1.4887	440
450	422.6	0.0194	1.0799	1.0993	430.1	774.5	1204.6	0.6280	0.8513	1.4793	450
460	466.9	0.0196	0.9748	0.9944	441.4	763.2	1204.6	0.6402	0.8298	1.4700	460
470	514.7	0.0198	0.9811	0.9009	452.8	751.5	1204.3	0.6523	0.8083	1.4606	470
480	566.1	0.0200	0.7972	0.8172	464.4	739.4	1203.7	0.6645	0.7868	1.4513	480
490	621.4	0.0202	0.7221	0.7423	476.0	726.8	1202.8	0.6766	0.7653	1.4419	490
500	680.8	0.0204	0.6545	0.6749	487.8	713.9	1201.7	0.6887	0.7438	1.4325	500
520	812.4	0.0209	0.5385	0.5594	511.9	686.4	1198.2	0.7130	0.7006	1.4136	520
540	962.5	0.0215	0.4434	0.4649	536.6	656.6	1193.2	0.7374	0.6568	1.3942	540
560	1133.1	0.0221	0.3647	0.3868	562.2	624.2	1186.4	0.7621	0.6121	1.3742	560
580	1325.8	0.0228	0.2989	0.3217	588.9	588.4	1177.3	0.7872	0.5659	1.3532	580
600	1542.9	0.0236	0.2432	0.2668	617.0	548.5	1165.3	0.8131	0.5176	1.3307	600
620	1736.6	0.0247	0.1955	0.2201	646.7	503.6	1150.3	0.8398	0.4664	1.3062	620
640	2059.7	0.0260	0.1638	0.1798	678.6	452.0	1130.5	0.8679	0.4110	1.2789	640
660	2365.4	0.0278	0.1165	0.1442	714.2	390.2	1104.4	0.8987	0.3485	1.2472	660
680	2708.1	0.0305	0.0810	0.1115	757.3	309.9	1067.2	0.9351	0.2719	1.2071	680
700	3093.7	0.0369	0.0392	0.0761	823.3	172.1	995.4	0.9905	0.1484	1.1389	700
705.4	3206.2	0.0503	0.0	0.053	902.7	0	902.7	1.0580	0	1.0580	705.4

Reproduced courtesy of GPSA after Wiley & Sons.

Vapor volume = 11,700 mol/D or 4.5 MMSCFD

Operating pressure = 10 psig

Operating temperature = 130 °F

Step 10. Lean Amine Cooler

Determine the lean amine cooler duty to cool the amine from the amine/amine exchanger outlet temperature of 168 °F to the contactor inlet temperature of 110 °F.

q = 293,000 lbs/h × 0.915 × (168 − 110)
= 15.50 MMBtu/h

Step 11. Booster Pump and Circulation Pump

Determine the lean amine booster pump and main circulation pump HP requirements.

$$BHP = \frac{(\Delta P)(L_{LA})}{1714e}$$

Assume P of 10 psi for lean/rich exchanger, 10 psi for amine cooler, 5 psi for filter, and 15 psi for associated piping.

$\Delta P = 2 \times 10 + 5 + 15 = 40$ psi

Assume e = 0.65

BHP = ((40)(560))/((1714)(0.65)) = 20.1 HP

For circulation pump, assume ΔP = (1000 psi)

$$BHP = \frac{(1000)(560)}{(1714)(0.65)}$$
= 503HP required for circulation pump.

DESIGN EXAMPLES (SI UNITS)

Example Problem 1: Iron Sponge Unit

Given:

Q_g = 2400 std m^3/h

SG = 0.6

H_2S = 19 ppm

P = 8400 kPa (A)

T = 38 °C

Mercaptans are not present

Solution:

Step 1. Calculate Minimum Vessel Diameter For Gas Velocity (Equation 1-3b)

SI units

$$d_{min} = 8.58\left(\frac{Q_g TZ}{PV_{g\,max}}\right)^{1/2}$$

Where

d_{min} = minimum internal vessel diameter, cm

Q_g = gas flow rate, std m³/h

T = operating temperature °K

Z = gas compressibility factor

= 0.85 (GPSA Figure 23-8)

(*Source:* Surface Production Operations, Volume 1, 3rd Edition, Figure 3-9)

P = Operating pressure, kPa

V_{gmax} = maximum gas velocity, m/s

Use $V_{gmax} = 3\ m/s$

$$d_{min} = 8.58\left[\frac{(2400)(311)(0.85)}{(8400)(3)}\right]^{1/2}$$

$d_{min} = 43.1$ cm

Step 2. Calculate Minimum Diameter For Deposition (Equation 1-4b)

SI units

$$d_{min} = 4255\left(\frac{Q_g X_{H_2S}}{\phi}\right)^{1/2}$$

Where

ϕ = rate of deposition g/h m² = 628

X_{H_2S} = mole fraction of H_2S = 19 ppm

$$d_{min} = 4255\left(\frac{(2400)(0.000019)}{628}\right)^{1/2}$$

= 36.3 cm

Step 3. Calculate Maximum Diameter (Equation 1-5b)

SI units

$$d_{max} = 8.58\left(\frac{Q_g TZ}{PV_{gmin}}\right)^{1/2}$$

Where

d_{max} = maximum internal vessel diameter, cm

V_{gmin} = minimum gas velocity, m/s

Use a V_{gmin} of 0.61 m/s

$$d_{max} = 8.58 \left(\frac{(2400)(311)(0.85)}{(8400)(0.61)} \right)^{1/2}$$

$$= 95.5 \text{ cm}$$

Therefore, any diameter from 43.1 to 95.5 cm is acceptable.

Step 4. Choose A Cycle Time of One Month or Longer (Equation 1-8b)

SI units

$$t_c = 1.48 \times 10^{-6} \frac{Fe\, d^2 He}{Q_g X_{H_2S}}$$

Where

t_c = cycle time, days

Fe = iron sponge content, kg Fe_2O_3/m^3

e = efficiency (0.65–0.8)

$$d^2H = \frac{t_c Q_g X_{H_2S}}{1.48 \times 10^{-6} Fee}$$

Assume $Fe = 116$ kg/m^3 and efficiency $= 0.65$

$$d^2H = \frac{(30)(2400)(0.000019)}{(1.48 \times 10^{-6})(116)(0.65)}$$

$$d^2H = 12,259$$

d (cm)	H (m)
55.9	3.9
60.96	3.3
66.0	3.1
71.1	2.4
76.2	2.1
91.4	1.5

An acceptable choice is a 76.2-cm vessel. Since t_c and e are arbitrary, a 3-m bed is appropriate.

Step 5. Calculate Volume of Iron Sponge to Purchase (Equation 1-7b)

SI units

$$Bu = 0.0022 \ d^2H$$

$$Bu_m = 7.85 \times 10^{-5} \ d^2H$$

Where

Bu = iron sponge volume, bushels

Bu_m = iron sponge volume, m^3

$$Bu = 0.0022(76.2)^2 \ (3)$$

$$Bu = 38 \text{ bushels}$$

Example Problem 2: Amine Processing Unit (DEA)

Given:

Gas volume	$= 120{,}000$ std m^3/h
Gas gravity	$= 0.67$ SG (Air $= 1.0$)
Pressure	$= 7000$ kPa (A)
Gas temperature	$= 38 \ °C$
CO_2 inlet	$= 4.03\%$
CO_2 outlet	$= 2\%$
H_2S inlet $= 19$ ppm	$= 0.0019\%$
H_2S outlet	$= 4$ ppm
Max. ambient temp	$= 38 \ °C$

Solution:

Step 1. Process Selection

Total acid gas inlet $= 4.03 + 0.0019$
$$= 4.032\%$$

Partial pressure of inlet acid gas $= 7000 \times (4.032/100) = 282$ kPa (41 psia)

Total acid gas outlet $= 2.0\%$

Partial pressure of outlet acid gas $7000 \times (2.0/100) = 140$ kPa (20 psia)

From 1 to 29 (CO_2 removal – no H_2S present) for removing CO_2 and H_2S, possible processes are Amines, Sulfinol®, and Carbonates.

The most common selection for this application is a DEA unit.

Step 2. DEA Circulation Rate (Equation 1-8b)

Determine the circulation rate.

Oilfield units

$$L_{DEA} = \frac{192 Q_g X_A}{c \rho A_L}$$

Where

L_{MEA} = MEA circulation rete, m^3/h

L_{DEA} = DEA circulation rete, m^3/h

Q_g = gas flow rate, std m^3/h

X_A = required reduction in total acid gas fraction, moles acid gas removed/ mole inlet gas. Note: X_A represents moles of all acid components, that is, CO_2, H_2S, and Meracaptans, as MEA and DEA are not selective

c = amine weight fraction, kg amine/kg solution (lbs amine/ lbs solution)

ρ = solution density, kg/m^3 (lbs/gal)

A_L = acid gas loading, mole acid gas/mole amine

= $1.045 \times 1000 \ kg/m^3 = 1045 \ kg/m^3$

$c = 0.35 \ kg/kg$

$A_L = 0.50 \ mol/mol$

$Q_g = 120,000 \ std \ m^3/h$

$X_A = 4.032\% = 0.04032$

Note: In order to meet the H_2S outlet, virtually all the CO_2 must be removed, as DEA is not selective for H_2S.

Determine the circulation rate

$$L_{DEA} = \frac{4.39(120,000)(0.04032)}{(0.35)(1045)(0.50)}$$

$$= 116 \ m^3/h$$

Add 10% for safety = 128 m^3/h

Step 3. Reboiler Duty (Equation 1-9b)

Determine the reboiler duty

SI units

$$q_{reb} = 77,421 L_{DEA}$$

Where

q_{reb} = reboiler duty, W

L_{MEA} = MEA circulation rete, m³/h (gpm)

L_{DEA} = DEA circulation rete, m³/h (gpm)

$$q_{reb} = 77,421(128)$$
$$q_{reb} = 10,000,000 \text{ W}$$

Step 4. Absorber Heat Balance

Perform a heat balance around the absorber as follows:

Set lean amine inlet temperature at 5 °C higher than inlet gas, or 43 °C.

Assume gas leaving absorber has 3 °C approach to inlet amine, or 46 °C exit temperature.

Determine heat of reaction for CO_2 absorbed using loading and circulation with safety factor considered.

Moles CO_2

$$\frac{\text{moles } CO_2}{\text{absorbed}} = \frac{120,000 \text{ std m}^3/\text{h} \times 0.04032}{10.87 \text{ std m}^3/\text{mol}}$$

$$= 445 \text{ mol/h } CO_2$$

$$\begin{aligned}\frac{\text{Moles DEA}}{\text{circulated}} = {}& 128 \text{ m}^3/\text{h} \times 1045 \text{ kg/m}^3 \times \\ & 0.35 \text{ kg/kg} \times \text{mol}/47.7 \text{ kg} \\ = {}& 980 \text{ mol/h}\end{aligned}$$

Loading moles CO_2/mole DEA = 445/980
$$= 0.45$$

From Table 1-5, using 35 wt% DEA, the heat of reaction for CO_2 is 1,395,000 J/kg

Heat of reaction due to the CO_2 is 1,395,000 J/kg × 445 mol/h × 1/3600 × 20 kg/mol = 3,450,000 W

Determine heat of reaction for H_2S absorbed.

$$\text{Moles } H_2S \atop \text{absorbed} = \frac{120{,}000 \text{ std m}^3/\text{h} \times}{10.87 \text{ std m}^3/\text{mol}} \frac{(0.000019 - 0.000004)}{}$$

$$= 0.166 \text{ mol h } H_2S$$

Use 1,300,000 Js/kg H_2S × 0.166 mol/h × 1/3600 × 15.4 kg/mol = 923 W

Calculate the DEA outlet temperature at the contactor.

$$\text{Gas flow rate} = \frac{120{,}000 \text{ std m}^3/\text{h} \times 0.67 \times}{1087 \text{ std m}^3/\text{mol}} \frac{13.14 \text{ kg/mol}}{}$$

$$= 97{,}200 \text{ kg/h}$$

Heat gained by gas stream

$$q = 97{,}200 \text{ kg/h} \times (46 - 38)\,°\text{C} \times$$

$$2700 \text{ J/kg}\,°\text{C} \times \frac{h}{3600\,s} = 580{,}000 \text{ W}$$

Heat lost to atmosphere—this depends on atmospheric temperature, surface area, wind velocity, etc. Assume 5% of reaction heat lost to atmosphere for an un-insulated absorber.

Total heat gained in outlet Amine = Heat of Reaction − heat gained by gas stream − heat lost to atmosphere.

Heat gained = 3,450,000 − 580,000 − (3,450,000 ×.05) = 2,700,000 W

Rich amine outlet temperature

$$= \frac{(2{,}700{,}000 \text{ J/s})(3600\,s/h)}{(128 \text{ m}^3/\text{h})(1045 \text{ kg/m}^3)} + 43\,°\text{C}$$
$$(0.915 \text{ Btu/lbs }°\text{F})$$
$$((4187 \text{ J/kg K})/(1 \text{ Btu/lbs }°\text{F}))$$

Step 5. Flash Tank

Determine flash tank size using the principles on Two Phase Separators, based on:

Operating pressure = 1035 kPa (G)

Operating temperature = 62 °C

Amine rate = 128 m³/h

$$(\text{Max}) \, CO_2 \text{ flashed} = 445 \text{ mol/h} \times \atop 20 \text{ kg/mol} \atop = 8900 \text{ kg/h}$$

(Max) H₂S absorbed = negligible

Size for 3 min retention time operating 1/2 full.

Step 6. Rich/Lean Exchanger

Determine the rich/lean amine exchanger duty:

Lean amine flow = 128 m³/h × 1045 kg/m³
= 134,000 kg/h

Rich amine flow = 134,000 + 445 mol CO_2/h × 20 kg/mol + 0.166 mol/h × 15.4 kg/mol
= 143,000 kg/h

Assume DEA Reboiler Temperature is 120 and 17 °C approach of Rich Amine to Lean Amine.

Rich Amine Inlet Temperature = 62 °C (from contactor)

Rich Amine Outlet Temperature = 120−17 °C = 103 °C

Calculate lean amine outlet temperature, assuming specific heat of rich amine is the same as the specific heat of lean amine = 3830 J/kg °K

$$T_{out} = 120\ °C - \left[(103 - 62) \times \frac{143,000}{134,000} \right]$$
$$= 76\ °C$$

Exchanger duty

$$= 143,000\ kg/h \times 3830\ J/kg\ °K \times (103 - 62)°K \times \frac{h}{3600\ s}$$

$$= 624 \times 10^6\ W$$

Step 7. Stripper Overhead

Set the stripper overhead condenser outlet temperature and calculate the flow rates of acid gas and steam. Size the condenser and determine the reflux rate. the acid gas and steam will be vented, flared, or processed further for removal of H₂S.

Set condenser temperature at 17 °C above maximum atmospheric temperature, that is, 55 °C.

From steam tables the partial pressure of water at 55 °C is 15.8 kPa.

Stripper and reflux drum will operate at 170 kPa (A).

Calculate the vapor rate leaving the reflux condenser

SI units

$$V_R = \frac{(P_R + 101.35)AG}{(P_R + 101.35) - PP_{H_2O}} \times \frac{1}{24}$$

Where

V_r = mole rate of vapor leaving condenser, kg mol/h

P_R = reflux drum pressure, kPa

A_G = moles acid gas/day, kg mol/day

PP_{H_2O} = partial pressure of water at the condenser outlet temperature, kPa abs

$$V_R = \frac{170 \times (mol\ CO_2 + mol\ H_2S)}{170 - PP_{H_2O}}$$

$$V_R = \frac{170 \times (445 + 0.166)}{170 - 15.8} = 491\ mol/h$$

$V_{steam} = 491 - 445.166 = 45.6\ mol/h$

$W_{H_2O} = 4.56\ mol/h \times 8.16\ kg/mol$
$= 372 kg/h$ lost from condenser

Step 8. Condenser Duty and Reflux Rate

Determine the condenser duty and reflux rate by performing an energy balance around the stripper.

$q_{reb} = q_{steam} + q_{H_2S} + q_{CO_2} + q_{amine} + q_{cond}$

The heat required to vaporize the acid gases (reverse the reaction) is

$q_{CO_2} = 3,450,000\ W$ and q_{H_2S}

Calculate the heat differential between the heat in and the heat out for the lean DEA

$q_{la} = 134,000\ kg/h \times$
$3830\ J/kg\ °K(120 - 103) \times \frac{1\ h}{3600\ s}$

$= 2,420,000\ W$

$$q_{cond} = 10,000,000 - 2,420,000$$
$$\qquad\qquad - 3,451,000$$

$$q_{cond} = 4,130,000 \text{ W}$$

The reflux condenser must cool the acid gas and steam from the top tray temperature to 55 °C and condense the amount required for reflux.

Assume a top tray temperature of 99 °C.

Heat required to cool the acid gas is approximately

$$q_{ag} = 445.2 \text{ mol/h} \times 20 \text{ kg/mol} \times$$
$$2700 \text{ J/kg °K} \times (99 - 55) \times \frac{1}{3600}$$

$$= 294,000 \text{ W}$$

$$W_{reflux} = \frac{(4,130,000 - 294,000) \text{ J/}s}{2676 - 230 \text{ J/g}}$$

$$= 1568 \text{ g/}s \frac{3600 \text{ }s}{h} = 5600 \text{ kg/h}$$

Calculate the vapors leaving the top tray

$$V_{top} = 445.2 \text{ mol/h AG} + \frac{(5600 + 372)}{8}$$

$$= 1200 \text{ mol/h vapors steam and acid gas}$$

$$PP_{H_2O} \text{ in top vapors} = \frac{(1200 - 445) \times 170 \text{ kPa}(A)}{1200 \text{ mol/h total}}$$

$$= 107 \text{ kPa}$$

This PP_{H_2O} of 107 kPa is equivalent to a temperature of about 103 °C from Table 1-7.

Recalculate, assuming top tray temperature of 103 °C. Then heat required to cool the acid gas is

$$q_{ag} = 445.2 \times 20 \times 2700 \times (103 - 55)$$
$$\times \frac{1}{3600}$$
$$= 321,000 \text{ W}$$

$$W_{reflux} = \frac{(4,130,000 - 321,000) \text{ W} \times \frac{3600}{1000}}{2680 - 230}$$
$$= 5600 \text{ kg/h}$$

$$V_{top} = 445.2 \text{ mol/h} + \frac{(5600 + 372)}{8}$$

$$= 1200 \text{ mol/h}$$

$$\text{mole/day water} = 1200 - 445.2 = 754.8$$

$$PP_{H_2O} \text{ in top vapors} = \frac{754.8}{1200} \times 170 = 107 \text{ kPa}$$

This is equivalent to a top tray temperature of 103 °C.

Step 9. Reflux Accumulator

Size the reflux accumulator using the principles on two phase separators for the following conditions:

> Liquid volume = 5600 kg/h water
>
> Vapor volume = 491 mol/h or 130,000 std m³/day
>
> Operating pressure = 170 kPa (A)
>
> Operating temperature = 55 °C

Step 10. Lean Amine Cooler

Determine the lean amine cooler duty to cool the amine from the amine/amine exchanger outlet temperature of 76 °C to the contactor inlet temperature of 43 °C.

$$q = 134,000 \text{ kg/h} \times 3830 \text{ J/kg K} \times$$
$$(76 - 43) \text{ K} \times \frac{1 \text{ h}}{3600 \text{ s}}$$

$$= 4,700,000 \text{ W}$$

Step 11. Booster Pump and Circulation Pump

Determine the lean amine booster pump and main circulation pump HP requirements.

$$\text{BHP} = \frac{(\Delta P)(L_{LA})}{3598e}$$

Assume P of 70 kPa for lean/rich exchanger, 70 kPa for amine cooler, 35 kPa for filter, and 100 kPa for associated piping.

$$\Delta P = 2 \times 70 + 35 + 100 = 275 \text{ kPa}$$

Assume $e = 0.65$.

$$\text{BHP} = \frac{(275)(128)}{(3598)(0.65)}$$
$$= 15 \text{ kW required for circulation pump.}$$

For circulation pump, assume $\Delta P = (6900 \text{ kPa})$.

$$BHP = \frac{(6900)(128)}{(3598)(0.65)}$$

$= 380 \text{ kW required for circulation pump.}$

NOMENCLATURE

A_L	acid gas loading, mole acid gas/mole amine
A_G	moles acid gas/day, kg mol/day (lbs mol/day)
BHP	pump brake horsepower, kW (HP)
Bu	iron sponge volume, m^3 (bushels)
c	amine weight fraction, kg amine/kg solution (lbs amine/lbs solution)
C_{PDEA}	DEA specific heat, kJ/kg K (Btu/lbs °F)
C_{PLA}	specific heat of lean amine, kJ/kg K (Btu/lbs °F)
C_{PMEA}	MEA specific heat, kJ/kg K (Btu/lbs °F)
D	vessel internal diameter, cm (in)
d_{max}	maximum internal vessel diameter, cm (in)
d_{min}	minimum internal vessel diameter, cm (in)
e	efficiency
Fe	iron sponge content, kg Fe_2O_3/m^3 (lbs Fe_2O_3/bushel)
H	bed height, m (ft)
h_L	enthalpy of water at condenser outlet temperature, J/kg (Btu/lbs)
h_s	enthalpy of steam at the top tray temperature, J/kg (Btu/lbs)
L_{DEA}	DEA circulation rate, m^3/h (gpm)
L_{H_2O}	water flow rate, m^3/h (gpm)
L_{LA}	lean amine circulation rate, m^3/h (gpm)
L_{MEA}	MEA circulation rate, m^3/h (gpm)
P	operating pressure, kPa absolute (psia)
P_R	reflux drum pressure, kPa, (psig)
PP_{H_2O}	partial pressure of water, kPa absolute (psia)
PP_i	partial pressure of component i, kPa absolute (psia)
q_{ag}	partial pressure of component i, kPa absolute (psia)
q_{cond}	condenser duty, W (Btu/h)
q_{cooler}	lean amine cooler duty, W (Btu/h)
q_{DEA}	DEA exchanger duty, W (Btu/h)
Q_g	gas flow rate, std m^3/h (MMSCFD)
q_{la}	lean amine solution heat duty, W (Btu/h)
q_{MEA}	MEA exchanger duty, W (Btu/h)
q_{ra}	rich amine solution heat duty, W (Btu/h)
q_{reb}	reboiler duty, W (Btu/h)
q_{vr}	heat duty to cool overhead vapors to condenser outlet temperature, W (Btu/h)
SG	specific gravity

SG_{DEA}	specific gravity of DEA
SG_{MEA}	specific gravity of MEA
T	operating temperature, K (°R)
t_c	cycle time, days
T_{in}	amine cooler inlet temperature, °C (°F)
T_{out}	amine cooler outlet temperature, °C (°F)
V_r	mole rate of vapor leaving condenser, kg mol/h (lbs mol/h)
W_{H_2O}	water flow rate, kg/h (lbs/h)
W_r	reflux rate, kg/h (lbs/h)
W_{steam}	steam rate overhead, kg/h (lbs/h)
V_g	gas velocity, m/s (ft/s)
V_{gmax}	maximum gas velocity, m/s (ft/s)
V_{gmin}	minimum gas velocity, m/s (ft/s)
X_A	required reduction in total acid gas fraction, moles acid gas removed/mole inlet gas
X_{H_2S}	mole fraction of H_2S
X_i	mole fraction of component i
Z	gas compressibility factor
ΔP	differential pressure, kPa (psi)
ΔT	temperature change, °C (°F)
λ	latent head of vaporization, J/kg (Btu/lbs)
ρ	solution density, kg/m^3 (lbs/gal)
ϕ	rate of deposition, grams/h m^2 (grains/min ft^2)

REFERENCES

1. Arnold, K. and Stewart, M.: "Surface Production Operations," Design of Gas-Handling Systems and Facilities," Gulf Publishing Co., (1995), Houston.
2. Personal communications with UOP and NATCO.

Part 2
Gas Processing

Contents

NGL RECOVERY CONSIDERATIONS

"Gas Processing" Is Used to Refer to the Removal of

> Ethane
>
> Propane
>
> *i*-Butane
>
> *n*-Butane

Liquids May Be

> Fractionated and sold as pure components.
>
> Combined and sold as natural gas liquids mix or NGLs mix.

Processing Objectives

> Produce transportable gas
>
> Meet sales-gas specifications
>
> Maximize liquid recovery

DOI: 10.1016/B978-1-85617-982-9.00001-6

Producing Transportable Gas

Remote locations require gas to be pipelined without condensation.

Condensation Has Two Drawbacks

Two-phase flow requires a larger pipe diameter than single-phase flow for the same ΔP.

When the two-phase stream arrives at its destination, elaborate slug-catchers may be required to produce the equipment downstream.

Two Alternatives Exist

NGL recovery at the remote site or

Dense-fluid pipelining

Meeting Sales-Gas Specifications

Most Gas Specifications Contain

Minimum gross heating value (GHV) specification and

Possibly a hydrocarbon (HC) dew point requirement.

If HC Is More Valuable as a Liquid

NGL removal should be maximized and

Still satisfy the minimum heating value specification.

If HC Is More Valuable as a Gas

It may be desirable to retain them as gas, subject to the HC dew point requirement.

Maximizing Liquid Recovery

A normal heating value specification of about 1000 Btu/scf can be met with methane alone as is shown in Table 2-1.

Gas streams containing N_2 and/or CO_2, which are incombustible, can require the presence of ethane to provide the required heating value.

Table 2-1 Typical hydrocarbon gross heating values

Hydrocarbon Component	GHV/(Btu/scf)
Methane	1010
Ethane	1770
Propane	2516

If the heavier HCs are more valuable as liquids, then complete liquification of propane and heavier HCs and partial ethane recovery is desirable.

Cycling of natural gas in a condensate reservoir, that is, reinjecting natural gas so as to keep the reservoir pressure above the gas dew point, will maximize ultimate NGL recovery. If the reservoir pressure is allowed to fall into the two-phase region, valuable liquids are condensed and will not be recovered.

VALUE OF NGL COMPONENTS

Ethane and Heavier HC Components (C_{2+}) Can Be Liquefied

Relative liquid and gas phase values of HCs are illustrated for propane.

In sales gas, propane is worth the contract price of the gas, assuming it can be left in the gas and sold for its GHV.

If natural gas is worth $5.00 MMBtu^{-1} and from the previous table the heating value of propane is 2516 Btu/scf, then

$(1,000,000$ Btu$) \times (1$ scf$/2516$ Btu$) = 397.5$ scf propane

If propane is liquified, the amount of liquid recovered at 60 °F is

$(397.5$ scf$) \times (1$ gal$/36.375$ scf$) = 10.9$ gal propane

Note: 36.375 scf/gal is read from the physical constant table under volume ratio scf gas/gal liquid.

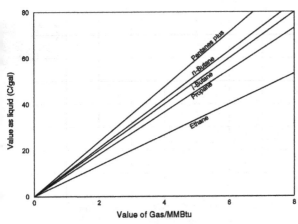

FIGURE 2-1 Energy equivalent of natural gas.

The equivalent value of this propane as a liquid is

$$\$5.00/10.9 \text{ gal} = \$0.459 \text{ gal}^{-1} \text{ (when gas is worth } \$5.00 \text{ MMBtu}^{-1})$$

If liquid propane can be sold for more than $0.459 gal^{-1} plus the cost of liquification, there is an economic incentive for propane liquification when gas is worth $5.00 MMBtu^{-1}.

Figure 2-1 shows the equivalent for NGL components as a function of gas price.

"Crude Spiking"

Increases the total barrels of oil

Raises the API gravity (increases the sales price per barrel).

Crude value is important

$0.459 gal^{-1} converts to $23.87 bbl^{-1}

Crude would have to be worth more than $23.87 bbl^{-1} to make the spiking economical.

The extent of condensate removal may be limited by the sales-gas GHV specification, particularly, if appreciable N_2 and/or CO_2 are present.

Figure 2-2 shows how ethane recovery is limited by inert gas content.

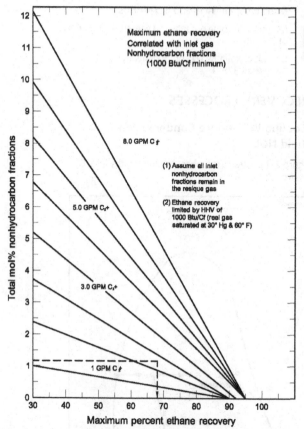

FIGURE 2-2 Maximum ethane recovery correlated with inlet gas nonhydrocarbon fractions.

GAS PROCESSING TERMINOLOGY

Definitions of LPG and NGL

Liquefied petroleum gas (LPG) products

Defined by their vapor pressure

Unfractionated NGL

Made up of pentanes and heavier HCs

May contain some butanes and very small amounts of propane

Cannot contain heavy components that boil at more than 375 °F

LIQUID RECOVERY PROCESSES

Any Cooling Will Induce Condensation and Yield NGL

Figure 2-3 illustrates the phase diagram paths for NGL recovery.

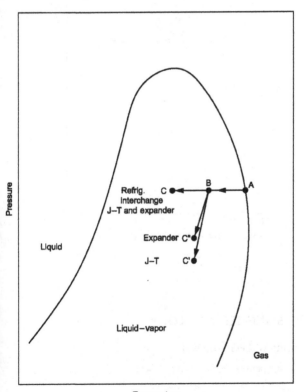

FIGURE 2-3 Phase diagram paths for NGL recovery.

The higher the pressure, the more condensation, other factors being equal.

Another NGL recovery technique is the use of a mass-transfer agent (MTA).

Basic NGL liquification processes are now described and related to Figure 2-3, when possible.

It is beyond the scope of this manual to discuss detailed designs of a gas processing plant.

Absorption/Lean Oil Process (Figure 2-4)

General Considerations

"Lean oil" (kerosene) is used to absorb light HC components from the gas.

Light components are separated from the rich oil and the lean oil is recycled.

Inlet gas is cooled by a heat exchanger with the outlet gas and a cooler before entering the absorber.

The absorber is a contact tower similar in design to a glycol contact tower.

Lean absorber oil trickles down over trays or packing while the gas flows upward.

Gas leaves the top of the absorber while the absorber oil, now rich in light HCs, leaves the bottom of the absorber.

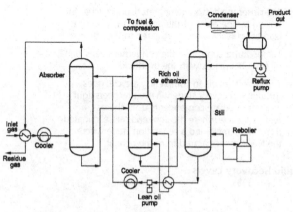

FIGURE 2-4 Simplified flow diagram of an absorption plant.

The cooler the inlet gas stream, the higher the percentage of HCs that will be removed by the oil.

Rich oil flows to the rich oil deethanizer (or demethanizer) to reject the methane or ethane (or methane alone) as flash gas.

In Most Lean Oil Plants

Rich Oil Dehydrator (ROD) unit rejects both methane and ethane because very little ethane is recovered by the lean oil.

If only methane were rejected, then it would be necessary to install a deethanizer column downstream of the still to make a separator ethane product and keep ethane from contaminating (increasing the vapor pressure of) the other liquid products made by the plant.

Heat is added at the bottom to drive off almost all the methane (and most ethane) from the bottoms product by exchanging heat with the hot lean oil coming from the still.

A reflux is provided by a small stream of cold lean oil injected at the top of the ROD.

Gas off the tower overhead is used as plant fuel and/or compressed.

Absorber oil then flows to a still where it is heated to a high enough temperature to drive off the propanes, butanes, pentanes, and other NGL components to the overhead.

The closer the bottom temperature approaches the boiling temperature of the lean oil, the purer the lean oil which will be recirculated to the absorber.

Temperature control on the condenser keeps lean oil from being lost with the overhead.

Thus, the lean oil, in completing a cycle, goes through a recovery stage where it recovers light and intermediate components from the gas, a rejection stage where the light ends are eliminated from the rich oil, and a separation stage where the NGLs are separated from the rich oil.

Liquid Recovery Levels

$C_3 = 80\%$

$C_4 = 90\%$

$C_{5+} = 98\%$

Disadvantages

Difficult to operate

Difficult to predict their efficiency at removing liquids from the gas as the lean oil deteriorates with time

Mechanical Refrigeration (Figures 2-5 and 2-6)

Supplied by a vapor compression cycle

Uses RF-22 as the refrigerant or working fluid

Inlet gas is cooled to a low enough temperature to condense the desired fraction of LPG and NGL.

Free water must be separated and the dew point of the gas lowered before cooling the feed to keep hydrates from forming.

TEG or molecular sieve or

Glycol injection

Glycol and water separate in the cold separator where they are routed to a regenerator, the water is boiled off and the glycol is circulated back to be injected into the inlet stream.

Ethylene glycol is used because of its low cost and at low temperatures it is not lost in the gas phase.

The chiller is typically a shell-and-tube, kettle-type exchanger.

RF-22 (which is cooled in a refrigeration cycle to −40 °F) is able to cool the gas to approximately −40 °F.

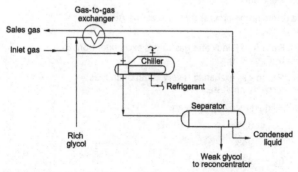

FIGURE 2-5 Simplified flow diagram of a refrigeration plant.

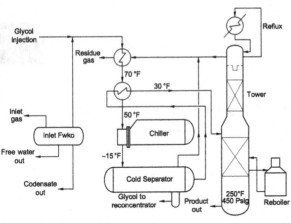

FIGURE 2-6 Mechanical refrigeration.

Gas and liquid are separated in the cold separator, which is a three-phase separator.

> Water and glycol come off the bottom.

> HC liquids are routed to the distillation tower.

> Gas flows out the top.

If it is desirable to recover ethane, this still is called a demethanizer.

If only propane and heavier components are to be recovered, it is called a deethanizer.

The gas is called "plant residue" and is the outlet gas from the plant.

The refrigeration process is shown as line ABC in the phase diagram.

> From A to B indicates gas-to-gas exchange; from B to C, chilling.

> Gas-to-gas exchange is very common in NGL recovery processes.

Typical liquid recovery levels are

$$C_3 = 85\%$$
$$C_4 = 94\%$$
$$C_{5+} = 98\%$$

These are higher than for a lean oil plant.

It is possible to recover a small percentage of ethane in a refrigeration plant.

Limited by the ability to cool the inlet stream to no lower than −40 °F with normal refrigerants.

Joule–Thomson (J–T) Expansions (Figure 2-7)

Inlet gas passes first through the gas-to-gas exchanger and then to an expansion or "choke" valve.

The expansion through the choke is essentially a constant enthalpy process.

Nonideal behavior of the inlet gas causes the temperature to fall with the pressure reduction, as shown by line ABC′ in the phase diagram of Figure 2-3.

The temperature change depends primarily on the pressure drop.

The J–T process is a "self-refrigeration" process, as opposed to mechanical refrigeration.

Again the condensed liquids must be fractionated to meet vapor pressure and composition specifications.

This process is most favored when the wellhead gas is produced at a very high pressure and can be expanded to sales line pressure with no recompression.

Cryogenic (Expansion Turbine) Plants (Figure 2-8)

Cool gas to −100 to −150 °F using expansion and J–T effect.

Chiller or J–T valve is replaced by an expansion turbine.

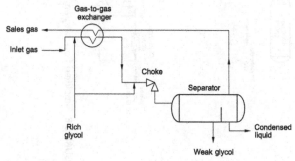

FIGURE 2-7 Low temperature separation (J–T valve).

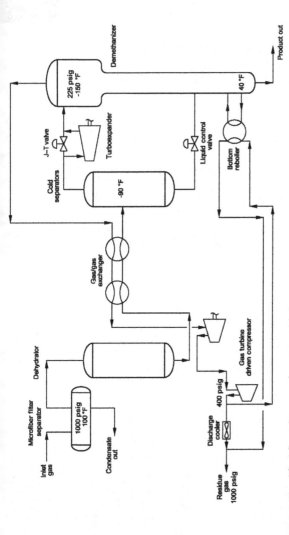

FIGURE 2-8 Simplified flow diagram of a cryogenic plant.

As the entering gas expands, it supplies work to the turbine shaft, thus reducing the gas enthalpy.

> Decrease in enthalpy causes a much larger temperature drop than that found in the J–T (constant enthalpy) process.

> Expansion process is indicated as line "ABC."

The gas is routed through heat exchangers where it is cooled by the residue gas, and condensed liquids are recovered in a cold separation at approximately −90 °F.

These liquids are injected into the demethanizer at a level where the temperature is approximately −90 °F.

Gas is then expanded (its pressure is decreased from inlet pressure to 225 psig) through an expansion valve or turboexpander.

Turboexpander uses the energy removed from the gas due to the pressure drop to drive a compressor, which helps recompress the gas to sales pressure.

The cold gas (150 °F) then enters the demethanizer column at a pressure and temperature condition where most of the ethanes-plus are in the liquid state.

As the liquid falls and is heated, the methane is boiled off and the liquid becomes leaner and leaner in methane.

Heat is added to the bottom of the tower using the hot discharge residue gas from the compressors to assure that the bottom liquids have an acceptable Reed Vapor Pressure (RVP) or methane content.

Because of the lower temperatures that are possible, cryogenic plants have the highest liquid recovery levels

> $C_2 = 60\%$

> $C_3 = 90\%$

> $C_{4+} = 100\%$

Advantages
> Simple to use

> Easy to package (more expensive than refrigeration)

PROCESS SELECTION

Normally, an economic comparison between viable alternatives will be required; however, the following guidelines are offered.

If the NGL Content of the Feed Gas Is Low

Use the expander process.

For Gases Very Rich in NGL

Simple refrigeration is probably best choice, while expansion is not usually satisfactory.

If the Inlet Gas Pressure Is Very High

Low temperature separation (ITS) may be attractive.

Low Inlet Gas Pressure

Favors an expander plant or straight refrigeration (if the gas is very rich).

Very Low Gas Rates

May justify only a very simple process such as an automatically operated J–T unit.

Large Flow Rates

Justify a more complex plant with more complex controls and more operating personnel.

Remote Wells

Remote wells may dictate simple operation and processing such as J–T plant.

A large number of wells may justify a central processing facility with more complex processing.

FRACTIONATION (FIGURE 2-9)

Bottoms liquid from any gas plant may be sold as a mixed product.

Bottoms liquid may be separated into its various components:

Ethane

Propane

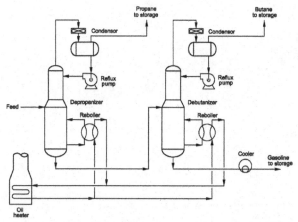

FIGURE 2-9 Simplified flow diagram of a fractionation system.

Butane

Natural gasoline

Fractionation splits liquids into its various components.

Liquids are cascaded through a series of distillation towers where successively heavier and heavier components are separated as overhead gas.

Specifications are normally in the form of RVP (controlled by upstream tower) and amount of heavy ends set by fractionator.

DESIGN CONSIDERATIONS

Proper design requires choosing an

Operating pressure

Bottoms temperature

Reflux condenser

Temperature

Number of trays.

Normally accomplished by using any one of several commercially available process simulation programs

Appendix A

Membrane/Amine Hybrid Grissik Gas Plant,[1,2,3] Sumatra, Indonesia: Case Study

Introduction

ConocoPhillips operates Grissik Gas Plant (Figure A-1) on behalf of its partners:

Talisman Energy

Pertamina

BPMigas

Design basis

Gas feed: 310 MMscfd

CO_2 concentrations

Inlet: 30%

Outlet: 3%

Process overview

CO_2 removal process uses a membrane/adsorption hybrid process

Utilizes both

Membrane separation and

Amine adsorption

Simplified process flow diagram is shown in Figure A-2.

Thermal swing adsorption (TSA) unit

Removes heavy hydrocarbons

Serves three functions

Membrane pretreatment

Feed gas dehydration

Sales gas hydrocarbon dew pointing.

Benefits of membrane/amine hybrid process

Have a single stage membrane and utilize the thermal value in the permeate stream, thereby

Enjoying the simplicity of a membrane separation process without the use of a recycle compressor while

Still avoiding hydrocarbon losses.

CO_2 rich permeate is sent to an atmospheric burner to produce steam which is used in the amine plant for regeneration.

Natural gas exiting the membrane

Contains about 15% CO_2

Fed to the amine absorption column where CO_2 is removed to about 3%.

Permeate rich in CO_2 exits the membrane at near atmospheric pressure.

Background

General Considerations

Plant built and commissioned in 1998 without TSA membrane pretreatment

Initial well tests indicated minimal amounts of heavy hydrocarbons

Subsequently, found not to be the case

DOI: 10.1016/B978-1-85617-982-9.00008-9

FIGURE A-1 Grissik gas plant.

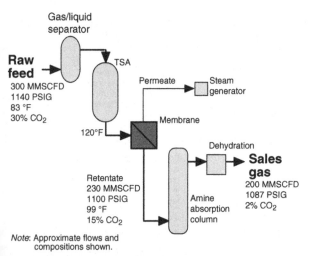

Note: Approximate flows and compositions shown.

FIGURE A-2 Grissik process flow diagram.

First Commissioning

Membrane initially installed with pretreatment consisting of

> Coalescing filter and
>
> Nonregenerable absorption guard bed.

At startup in 1998

> Actual levels of heavy hydrocarbons (CO_{10+}, aromatics, and napthenes) were higher than anticipated.

Resulted in sharp reduction in membrane capacity (declining to 20–30% of initial capacity within in a month)

To maintain production capacity, the membrane elements were being frequently replaced.

Installation of TSA

> ConocoPhillips evaluated heavy hydrocarbon removal processes including

Gas chilling process and

Regenerable adsorption process.

Gas chilling process

Deemed ineffective at the plant operating pressure, which was near the cricondenbar of the feed gas phase envelope

Regenerable adsorption process

Short cycle process from Engelhard which uses Sorbead (Silica Gel) as the adsorbent

Uses multiple beds in parallel adsorption to remove

Heavy hydrocarbons

Aromatics

Napthenes

Adsorption cycle is followed by regeneration of the silica gel at elevated temperatures.

TSA was built and installed by Kvaener in 2000

Designed to reduce C_{6+} components (including aromatics and napthenes) so that membrane performance can be maintained for an extended period of time

Designed with two separate trains, each with four adsorption vessels (refer to Figure A-3)

TSA design and performance

General Design Considerations

Since feed gas was found to contain high levels of heavy hydrocarbons (C_{10+}, aromatics, and napthenes)

TSA had two functions and solved two problems:

TSA removes heavy hydrocarbons for proper pretreatment so as to yield long membrane life.

FIGURE A-3 Engelhard thermal swing adsorption unit.

Removal of heavy hydrocarbons allows the sales gas to meet hydrocarbon dew point specs.

Since water is more strongly held onto the Sorbead adsorbent than any of the hydrocarbons, the TSA system also dehydrates the feed upstream of the membrane unit.

TSA Process Description

Each train was designed to treat 225 MMscfd

Consists of four internally insulated adsorber towers

Minimize the thermal mass for the short thermal cycle

Reduces heat load on the system

Refer to TSA process flow diagram (Figure A-4)

Feed gas, after passing through the two-phase separator, is split into two parallel paths.

Majority of the gas flows through the pressure drop valve and then directly to two towers on parallel adsorption.

Cycle time of the towers is staggered by 50% to allow for a continuous flow of

treated gas to the downstream membrane unit.

Balance of the feed gas bypasses the pressure drop valve so as to provide the necessary flow through the towers being cooled and heated.

Regeneration path contains the

Tower being cooled

Regeneration heater

Tower being heated

Heat recovery heat exchangers, and

Spent regeneration gas separator

Each tower is associated with six valves that allow it to change functional positions

Adsorbing

Heating/regenerating

Cooling

Adsorbing

Wet feed gas is used as the regeneration medium, and because of the pressure drop valve, there is no need for a compressor to boost the pressure of the spent regeneration gas.

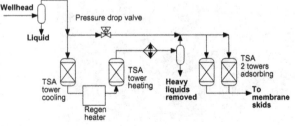

FIGURE A-4 TSA process flow diagram.

During adsorption

> Water and C_{6+} components are adsorbed at 1100 psig and 90–140 °F.

> Prior to C_{6+} breakthrough, the tower position is switched to heating mode and is completely heated to 540 °F.

> Internal insulation allows heating of the adsorbent only and not the steel shell.

> During heating, the water vapor and C_{6+} components are desorbed

Spent regeneration gas stream containing water and C_{6+} is then cooled and

> Condensed liquids removed in the regeneration gas separator.

> This is the only place in the process where the heavy hydrocarbons exit the system.

Reasons for Four Towers

In order to maintain an acceptable flow velocity across the adsorber bed, the number of towers used is a function of

> Flow rate and

> Tower diameter.

Maximum tower diameter was determined by transport limits; Grissik design resulted in four towers with two towers in parallel adsorption.

> Internal insulation was used to minimize the amount of regeneration gas requirement.

Heating and cooling towers are in a series arrangement which also conserves the amount of regeneration gas required.

Additional benefit of having towers in parallel on adsorption is an equalized composition of the treated gas.

In a single tower system

> There is a difference in the gas composition between beginning and end cycle, caused by the breakout of the individual components.

In a four tower system with two towers on adsorption

> There is an offset time of half an adsorption cycle.

> Gas composition of the combined outlet gas is more constant than from a single tower system.

Cycle Times and Breakthrough

Cycle times

> Driven by the breakthrough behavior of the C_{6+} components in the tower design in order to meet the hydrocarbon specification of the treated gas

> Result of analysis and field observations

> Typical cycle consisted of (Refer to Table A-1)

>> Two-hour adsorbing

>> One-hour heating, and

>> One-hour cooling.

Heat Recovery Between Cooling and Heating

System uses one tower heating and one tower cooling at a time.

Table A-1 Tower mode timing

Tower 1	2 h Adsorption	1 h Heating	1 h Cooling
Tower 2	1 h Cooling	2 h Adsorption	1 h Heating
Tower 3	1 h Heating	1 h Cooling	2 h Adsorption
Tower 4	1 h Adsorption	1 h Heating	1 h Cooling, 1 h adsorption

Hot gas leaving the tower being cooled flows through the heater in order to get additional heat in.

At the beginning of the cycle, gas exiting the tower on cooling is almost at the required heating temperature.

Results in nearly no make-up heat being required

Due to the entire tower being cooled the gas is at the hot regeneration temperature of 540 °F

During the cooling cycle, the temperature of the gas exiting the cooling tower

Drops so the heater has to provide the required heating gas temperature.

Gas-to-gas heat exchanger

It is used to capture the heat exiting the tower which is being heated.

Hot gas is cross-exchanged with the gas upstream of the regeneration gas heater (Refer to Figure A-4).

Exchanger is bypassed during the time when the gas exiting the tower on cooling is at a higher temperature than the gas leaving the tower in the heating step.

Regeneration heater

Direct-fired heaters

Size of the heater depends on the regeneration gas flow required to heat the adsorption bed and desorb the water and hydrocarbons within the design cycle time.

TSA performance

After recommissioning the plant in October 2000

Good TSA performance removing the heavy hydrocarbons led to excellent membrane performance.

TSA performance regarding hydrocarbon dew point was impressive, see Table A-2.

Corresponding phase envelopes are shown in Figure A-5.

Figure A-6 shows the results of gas sampling done with a mass spectrometer where both the feed and exit streams of the TSA were analyzed dynamically.

Ratio of hydrocarbon concentration in the outlet versus inlet is shown.

Note the strong cutoff that occurs between C_6 and C_8.

Heavier hydrocarbons are essentially completely removed.

Table A-2 TSA hydrocarbon dew point

TSA feed gas	86 °F at 1150 psig
TSA outlet gas	−22 °F at 1115 psig

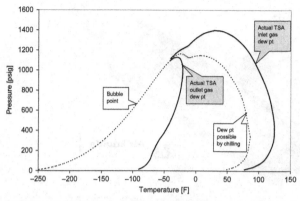

FIGURE A-5 TSA inlet and outlet phase envelopes.

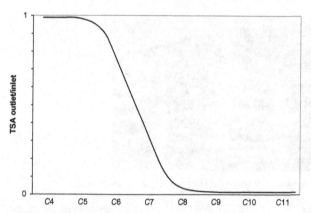

FIGURE A-6 TSA hydrocarbon tail.

Air liquide—medal membrane

General Considerations

Polyimide hollow fiber membrane elements (shown in Figure A-7) provide for a high efficiency separation of CO_2 from hydrocarbon streams.

Membrane system was fabricated as multiple skids (refer to Figure A-8) operating in parallel.

Each skid contains multiple horizontal tubes.

Each tube contains multiple membrane elements (refer to Figure A-9).

> Multiple elements are installed in a single tube.
>
> Membrane elements are actually functioning in parallel.

More than 100 membrane elements are used in this plant.

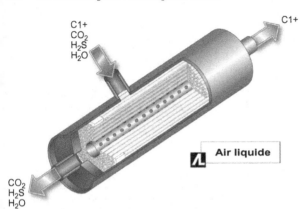

FIGURE A-7 Air Liquide—MEDAL natural gas membrane.

FIGURE A-8 Skid containing Air Liquide membrane elements.

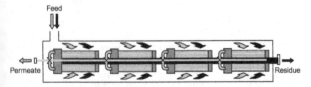

FIGURE A-9 Multiple membrane element flow arrangement.

Feed gas enters the tube near one end and flows axially to all the membrane elements by way of an annular clearance.

Each element is composed of several hundred thousand parallel hollow polyimide fibers.

Feed gas enters the membrane elements on the fiber shell inside and flows over the fibers, where CO_2 is removed, to a coaxial tube in the center of each element (retentate).

Retentate streams for each element flow axially to exit at one end of the tube.

CO_2 selectively permeates into the bore of the fibers and then flows axially to a collection point at the end of each element (permeate).

Permeate of each element is then collected in the coaxial center tube and flows axially to exit the tube at the opposite end from the retentate.

Membrane Performance

Typical operating conditions

Membrane skids are fed directly from the output of the TSA.

Feed temperatures vary between 90 and 120 °F.

Feed pressure is 1100 psig.

Feed gas contained 30% CO_2.

Permeate pressure is about 10 psig which flows to the steam generator burners.

Hydrocarbon losses versus time

One of the major advantages of the polyimide membrane is its ability to maintain integrity indefinitely, even aging in the presence of heavy hydrocarbons.

As shown in Figure A-10, membrane integrity is solid and the hydrocarbon losses have decreased somewhat since startup.

This trend of decreasing hydrocarbon losses indicates no loss of membrane integrity and actually shows a slight increase in apparent intrinsic membrane selectivity.

Such a selectivity increase would be consistent with the change in permeability (see below).

Membrane capacity versus time

After TSA was commissioned in October 2000

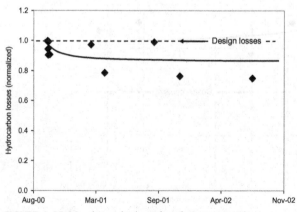

FIGURE A-10 Membrane hydrocarbon losses versus time.

One membrane skid was retrofitted with new membrane elements and its performance tracked.

Results shown in Figure A-11

Vertical axis is labeled "Relative Capacity to Remove Moles of CO_2" which is the normalized intrinsic membrane permeability.

Initial capacity was well above design, and after 10 years of operation, the capacity still remains above design.

Exact membrane life can be extrapolated to be over 12 years without replacement

Excellent operation of the TSA and membrane have resulted in years of trouble free operation with zero maintenance, that is, no membrane replacements.

Membrane skids were shut down and restarted many times for maintenance of surrounding equipment or capacity turndown.

Start and stop, or pressurization and depressurization cycles have no effect on membrane performance, although caution must be used to avoid reverse pressurization.

Permeate/Acid Gas Utilization

Two waste heat boiler units are installed.

Waste heat boilers recover "waste heat" available in low BTU permeate gas stream (150–250 Btu/scf) from the membrane units.

Utilizing waste heat in the permeate stream means single stage membrane can

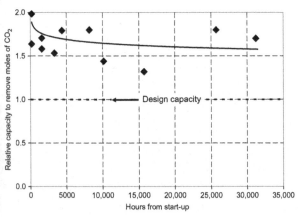

FIGURE A-11 Membrane capacity versus time.

be used without the limitations of a second membrane stage with the accompanying recycle gas compressor and still avoid hydrocarbon losses.

Boilers are designed to "incinerate" the acid gases removed by the amine unit.

Auxiliary fuel is utilized to make up any inadequacy of heating value input and to stabilize the flame.

Furnace temperature is maintained above 1600 °F prior to introducing permeate fuel or acid gases.

Lower temperature leads to incomplete destruction of the component and results in the emission hazards.

Waste heat boilers are controlled by steam header that actuates pressure control valves on each steam drum.

Output of the steam header pressure controller goes through flow ratio controllers of permeate gas, fuel gas, and combustion air.

Fuel gas flow rate is set around 10% of permeate gas flow rate while combustion air is controlled to ensure stoichiometry and complete combustion with 2–5% excess air.

Waste heat boiler produces steam up to 210,000 lbs/h at 150 psig and 348 °F.

Biggest consumer of steam produced is the amine system.

Condensing heat released by the steam is used to remove acid gas from amine solvent at amine reboilers.

Amine System

Amine system further reduces CO_2 and H_2S to meet sales gas specification.

Residue gas from the membrane unit, containing 15% CO_2,

> Flows into the amine contactors and
>
> Contacted with lean amine (50%wt-activated MDEA).

CO_2 absorption by activated MDEA is limited to a maximum loading of 0.5 mol acid gas/mol MDEA.

CO_2 content in the treated gas varies between 2% and 5% by volume (3%-vol average).

Rich amine is then flashed at 75 psig, heated through a lean/rich amine exchanger, and regenerated by the steam heated reboiler.

The 150 psig steam used for regenerating amine is produced in the waste heat boiler that burns permeate gas.

Several common problems of an amine system include

> Reduced strength and ability to absorb acid gas
>
> Degradation
>
> Foaming and
>
> CO_2 corrosion attack during acid gas breakout inside the reboiler.

Most problems found in an amine system are due to the presence of contaminant in the amine solvent, including

> Heat stable salts
>
> Degradation products
>
> Injected chemicals
>
> Hydrocarbons and
>
> Particulates.

Heat stable salts and degradation products are formed by amine solvents that decompose and/or react with other contaminants.

TSA/membrane installed upstream of the amine system has mitigated the above problems to an acceptable level.

> TSA unit removes heavy hydrocarbons from the feed gas and nearly eliminated the foaming risk of amine solvent.

> An antifoam injection system is provided to anticipate worst case conditions.

CO_2 content reduction by the membrane unit

> Breakout in the regeneration process

> Lessons CO_2 breakout in the regeneration process

> Reduces contaminants that may trigger salt formation or amine degradation

> Though contaminants could also be introduced by makeup water or even makeup amine

Treated gas condition at the outlet of the amine unit is normally 3%-vol CO_2 and 2–4 ppmv H_2S, while sales gas contract specifies 5%-vol CO_2 and 8 ppmv H_2S.

One advantage from the high performance of absorption is an allowance to increase system deliverability by bypassing some untreated gas and blending with treated gas while maintaining the sales gas specification.

References

1. Anderson, C.L. and Siahaan, A. "Case study: Membrane CO_2 removal from natural gas, Grissik gas plant, Sumatra, Indonesia", Regional Symposium on Membrane Science and Technology, 2004, Johor Bahru, Malaysia.
2. Malcolm, J. "The Grissik gas plant", Hydrocarbon Asia, 2001.
3. Anderson, C.L. "Case study: Membrane CO_2 removal from natural gas", Regional Symposium on Membrane Science and Technology, 2004, Johor Bahru, Malaysia.

Appendix B

Judge Digby Gas Plant Hikes Production with Quick Solvent Change-Out[1]: Case Study

Judge digby plant

The gas plant is operated by BP and is located in Pointe Coupee Parish in South Louisiana.

Plant came online in 1970.

Inlet feed stream composition (Table B-1)

DEA amine unit produces a sales gas with less than 3% CO_2 and 8 ppm H_2S.

Consists of two trains

    Train 1 consists of 150 MMcfd conventional amine unit using 30% DEA followed by a TEG dehydration unit.

    Train 2 consists of 100 MMcfd TEG dehydration unit only.

Figure B-1 is an aerial view of the plant.

    Train 1 is shown in the upper left-hand corner.

Figure B-2 is the process flow diagram of Train 1.

    Plant was initially designed to produce 150 MMcfd of treated gas.

    Train 1 was designed to remove 95% of acid gas from 120 MMcfd raw gas in the sweetening unit.

    30 MMcfd stream of raw gas bypasses the amine unit and combines with amine plant-treated gas to yield 150 MMcfd that meets the desired CO_2 and H_2S specification.

    Bypass gas is treated with a nonrecoverable H_2S scavenging chemical before being blended with the sales gas.

In 1999 engineering study indicated

    Maximum throughput was between 135 and 140 MMcfd capacity declined to 135 MMcfd.

    Unit became unstable and required 24-h manned operation at production rates greater than 140 MMcfd.

DEA system exhibited several problems.

    DEA solution had degraded and the reboilers were severely fouled.

    Corrosion probes indicated a high degree of corrosion.

    Regenerator still was hydraulically limited and unable to fully strip the rich DEA solution.

    Lean solution loadings of 0.6–0.7 mol of CO_2/mol of DEA were typical.

        Fully stripped DEA should contain less than 0.02 mol of CO_2/mol of DEA.

        High solution lean loadings and solution

DOI: 10.1016/B978-1-85617-982-9.00009-0

Table B-1 Typical Inlet Feed Stream.

Component	Mole percent
CO_2	8.08
C_1	89.54
C_2	1.15
C_3	0.18
$i\text{-}C_4$	0.08
$n\text{-}C_4$	0.05
$i\text{-}C_5$	0.05
$n\text{-}C_5$	0.03
C_{6+}	0.57
N_2	0.27
H_2S ppm	40

degradation products usually lead to corrosion and reboiler fouling problems.

High CO_2 loading in the DEA from the regenerator also prevented the absorber from removing all the CO_2 from the raw gas; therefore, there was too much CO_2 in the gas leaving the DEA unit.

Instead of treating and bypassing 122.5 and 30 MMcfd, respectively, the plant could only treat and

FIGURE B-1 Judge Digby gas processing plant.

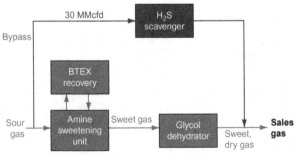

FIGURE B-2 Judge Digby flow diagram.

Table B-2 Operating history with DEA

	August 1999	July 2000
Inlet gas flow rate (MMcfd)	92.5	135.0
Bypass gas flow rate (MMcfd)	0	13.5
Gas flow rate to absorber (MMcfd)	92.5	122.5
Inlet gas pressure (psig)	994	1000
Inlet gas temperature (°F)	96	95
DEA circulation rate (gpm)	680	953
DEA concentration (wt%)	30	33
Lean DEA loading (mol CO_2/mol DEA)	0.02	0.06
Lean DEA temperature (°F)	113	110
Reboiler duty (MMbtu/h)	39.5	50
CO_2 in outlet gas (%)	2.72	2.75

bypass 122.5 and 13.5 MMcfd, respectively, to produce a combined sales gas volume of only 135 MMCFD that contained less than 3% CO_2.

Table B-2 shows the plants' performance using DEA.

Debottlenecking

BP considered several options to regain lost capacity.

Contactor was hydraulically limited to a feed rate of 120 MMcfd, which also limited BP's options.

Flow rates greater than the contactor's hydraulic limit resulted in

Large amine losses

Increased corrosion, and

Operating instabilities.

To maximize production, BP wanted to maximize

Bypass gas flow rate and

Amount of CO_2 removed in the absorber.

BP also wanted to reduce the number of plant upsets at high flow rates.

Plant is unmanned 16 h/day.

Upsets during unmanned periods increase the number

and cost of off-shift operator callouts.

Each unplanned shutdown had adverse effects on producing wells.

BP decided to replace the existing DEA chemical solvent with Dow's AP-814 solvent.

Requires less regeneration duty

Absorbs more CO_2

Solvent change-out required

Less than 24 h without extensive system cleanout and

No mechanical equipment modifications.

Preparing for the conversion

Figure B-3 shows the

Amine contactor (background)

Regeneration still, and

Benzene–toluene–ethylbenzene–xylene (BTEX) stripper (foreground).

Figure B-4 shows a process flow diagram of the Train 1 amine system.

BP performed a gamma scan of the absorber and regenerator before the solvent switch.

FIGURE B-3 Amine Sweetener includes a contactor, regenerator and a BETEX stripper.

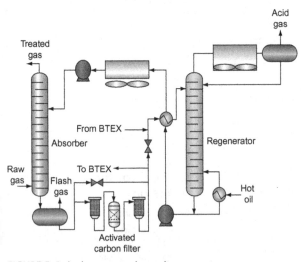

FIGURE B-4 Amine sweetening unit.

An engineering analysis of the reboiler showed it was severely fouled.

Two cleaning crews manned the shutdown to ensure timely cleaning of the reboiler tubes.

Cleaning process was the time-limiting step.

32 h were spent cleaning the severely fouled tubes before halting the process.

DEA was removed from the storage tank and shipped to another plant.

The storage tank was cleaned and filled with an initial charge of 45 wt% AP-814.

Two antifoams were brought to the plant:

> One antifoam used for start-up that would prevent foaming created by solids and

> One antifoam used for normal operations.

The initial plant configuration did not allow continuous antifoam addition to the stripper; thus an antifoam charge pump was added.

Process Safety Management considerations

> A Management of Change (MOC) document was developed and

> Process Hazards Analysis review was performed.

>> Considered all the changes needed to accommodate a solvent switch, including special issues such as

>>> Metallurgy

>>> Equipment configuration

>>> Pump design and

>>> Gasket materials

>> Review indicated that no additional modifications were required.

Training

> Conducted for

>> Unit operators and

>> Company engineers.

Focused on

> New laboratory test methods

> Operating techniques and

> Operating parameters with the new AP-814 solvent.

Operations designed a time window for the reboiler cleaning and solvent change-out.

The turnaround

In October 2000

> DEA was removed from the sweetening unit and drained from all the low points as much as possible.

> Amine unit was flushed with water drained.

> Activated carbon filter was emptied and filled with a fresh charge.

After start-up, a gas chromatographic analysis indicated that only trace amounts of DEA remained in the system.

Solvent swap

> Initially, inlet and outlet isolation valves were to isolate the reboilers.

> Isolation valves leaked and reboilers had to be cleaned before fresh solvent was added to the unit.

> Solvent swap would have taken 4 h if the isolation valves did not leak.

Reboiler tube fouling

> Reboiler tubes were hydroblasted in an attempt to remove the iron carbonate fouling.

After 32 h hydroblasting was stopped and the reboiler was put into service.

20–30% of the tubes remained plugged with the iron carbonate.

Unit started up smoothly with the new solvent.

Plant produced the expected additional 20 MMcfd of gas, even with the fouled reboiler tubes.

Plant operations

Plant operated for 6 weeks when well production was lost.

Table B-3 shows performance data collected during that time.

Since the new solvent removed more CO_2, sweetened gas could be mixed with untreated gas.

New solvent allowed more operating flexibility, which helped compensate for less than optimum reboiler performance.

Damaged wells were restarted in March 2001.

Plant could not run at high rates because the reboiler tube condition had worsened.

Reboiler tube bundle was replaced.

Table B-4 shows the plant performance with AP-814 and the historical maximum performance with DEA.

After bundle replacement

Plant had no problem meeting design capacity.

At $3.00 Mcf^{-1}, this translates to an incremental increase in sales revenues of approximately $34 million/year.

Table B-3 Operating performance with Ucarsol

Inlet gas flow rate (MMcfd)	131
Bypass gas flow rate (MMcfd)	26
Gas flow rate to absorber (MMcfd)	105
Inlet gas pressure (psig)	1001
Inlet gas temperature (°F)	102
Solvent circulation rate (gpm)	840
Lean solvent loading (mol CO_2/mol solvent)	0.03
Lean solvent temperature (°F)	117
Reboiler duty (MMbtu/h)	48
CO_2 in outlet gas (%)	0.01

Table B-4 Optimized performance comparison

	DEA (33 wt%)	Ucarsol (45 wt%)
Maximum processing capacity (MMcfd)	136	150
Solvent circulation rate (gpm)	953	1000
Solvent CO_2 loading (mol CO_2/mol solvent)		
Lean	0.06	0.03
Rich	0.49	0.46
Reboiler duty (MMbtu/h)	50	50
CO_2 in treated gas		
Outlet from contactor (%)	2	0.2–0.5
Sales gas with bypass (%)	3	<2.5

Plant performance

OPEX remained the same with the AP-814 despite the capacity increase.

Amine unit could theoretically process up to 168 MMcfd if other bottlenecks were removed.

Lab tests did not show any solvent degradation.

Solvent losses were low.

Sweetening unit ran for months without any significant change in solvent concentration.

During the change-out, the 10-µm particulate filters were replaced with 5-µm filters.

New 5-µm filters are changed-out half as often.

Corrosion is low.

Corrosion monitoring probes that indicated high levels of corrosion with DEA are now reading low levels.

Low iron levels in the solvent and good performance from the 5-µm filters confirm this finding.

Sweetening unit performed well during upstream upsets and well outages.

Betex emissions

Judge Digby gas stream contains some BETEX.

BETEX compounds have a high fuel value.

Regulated for human contact and air emissions.

BETEX is more soluble in amine solvents than other hydrocarbons.

Plant has a BETEX-removal unit that strips BETEX from the rich amine.

Unit is on the rich-amine line between the rich-amine filters and lean-rich heat exchanger.

Fuel gas is the stripping agent.

This gas, along with any gas stripped out with the BETEX, is recovered and routed to the plant's fuel-gas system.

The BETX unit was evaluated using AP-814 solvent at a higher circulation rate. The evaluation indicated the change from DEA did not significantly alter the load on the BETEX-removal system.

Reference

1. Hlozek, R., & Jackson, S. "Louisiana gas plant hikes production with quick solvent changeout", Oil and Gas Journal, June 9, 2003.

INDEX

Note: Page numbers followed by *f* indicate figures and *t* indicate tables.

ted in the United States
Bookmasters